ESSENTIAL MATH FOR CHEMISTRY STUDENTS

David W. Ball

Associate Professor of Chemistry
Cleveland State University
Cleveland, OH

BROOKS/COLE

™

THOMSON LEARNING

Australia • Canada • Mexico • Singapore • Spain • United Kingdom • United States

No writing project is the effort of just one person. Several other people have been instrumental in the development of this project, and they all have my thanks. Keith Dodson of West Educational Publishing Company provided timely feedback and guidance. Rick Mixter, also from West, and Jennifer Welchans and Trent Carruthers, the local West representatives, were very supportive. Helen Newman was my first "guinea pig" and agreed to read the entire manuscript from cover to cover, worked the problems, and gave me an honest appraisal of the manuscript. Students Michelle Ballash, David Fuote, Sheri Salay, and Erine Stames class–tested the manuscript under real classroom conditions and made some invaluable suggestions.

WEST'S COMMITMENT TO THE ENVIRONMENT

In 1906, West Publishing Company began recycling materials left over from the production of books. This began a tradition of efficient and responsible use of resources. Today, 100% of our legal bound volumes are printed on acid-free, recycled paper consisting of 50% new paper pulp and 50% paper that has undergone a de-inking process. We also use vegetable-based inks to print all of our books. West recycles nearly 27,700,000 pounds of scrap paper annually—the equivalent of 229,300 trees. Since the 1960s, West has devised ways to capture and recycle waste inks, solvents, oils, and vapors created in the printing process. We also recycle plastics of all kinds, wood, glass, corrugated cardboard, and batteries, and have eliminated the use of polystyrene book packaging. We at West are proud of the longevity and the scope of our commitment to the environment.

West pocket parts and advance sheets are printed on recyclable paper and can be collected and recycled with newspapers. Staples do not have to be removed. Bound volumes can be recycled after removing the cover.

Production, Prepress, Printing and Binding by West Publishing Company.

 TEXT IS PRINTED ON 10% POST CONSUMER RECYCLED PAPER   ∞

Copyright © 1996 Wadsworth Group.
Brooks/Cole is an imprint of the Wadsworth Group,
a division of Thomson Learning, Inc.
Thomson Learning™ is a trademark
used herein under license.

Wadsworth Group/Thomson Learning
10 Davis Drive
Belmont CA 94002-3098
USA

For information about our products, contact us:
Thomson Learning Academic Resource Center
1-800-423-0563
http://www.wadsworth.com

For permission to use material from this text, contact us by
Web: http://www.thomsonrights.com
Fax: 1-800-730-2215
Phone: 1-800-730-2214

Printed in the United States of America
10 9 8 7 6 5

ISBN 0–314–09604–3

Contents

iii

Preface

To the Instructor

In my experience, many students come into their first chemistry sequence – either at the allied health or at the science/engineering level – with a relative inability to perform the basic mathematical skills that are necessary to do some of the material. Every year, I see students dropping general chemistry not because they don't understand the **chemistry**, but because they can't do the **math**. Some of them can't even use their own calculator correctly, even though the students think they can. Many of these students say that they have satisfied the math prerequisite for chemistry, which is usually high-school algebra. However, when asked to actually perform math (like on an exam), they come up short.

This book is meant to serve as a review of essential math skills for the chemistry student. Although most students have had algebra, it may have been so long ago or the knowledge may not have been used since that time. Many students have deficient math skills and are not truly prepared for chemistry.

This book will help resharpen the math skills that are necessary for most introductory chemistry courses. It is not written to accompany any specific chemistry textbook, and should even be considered a stand-alone book. It is an appropriate supplement for chemistry textbooks on the preparatory, allied health/general-organic-biological (GOB), and science/engineering general chemistry level. It does not cover trigonometry or calculus, since most general chemistry texts do not use these topics directly. (Indeed, there are only a few places where it is used, but mostly it seems applied in a very "hand-waving" fashion.) Whether these topics will be included in future editions depends on how their absence is received by professors and students alike.

Each chapter has exercises worked out in detail and student exercises at the end. Answers to all of the student exercises are included with each chapter, but are intentionally on the left page so they aren't exposed unintentionally. Material that discusses calculator use has calculator key sequences in boldface and sans serif font, to distinguish it from the rest of the material. The manuscript was class-tested by volunteer students in one of our first–term general chemistry courses.

The intended size of this book means that a lot of specific types of chemistry problems could not be covered directly. For example, percent yield calculations and limiting reagent problems are not discussed, nor are all of the huge variety and complexity of equilibrium constant problems. These problems can be found in a chemistry book. Hopefully, this book will help provide the confidence in the math skills that will be necessary to work these types of problems successfully.

To the User

Good math skills are essential in order to master the basics of any science, including chemistry. If you think that you are deficient in those skills, then this book should help you brush up on the necessary material you will need to learn chemistry.

You are encouraged to start at page 1 and go through **the entire book**, a page at a time. The later chapters build on material of the previous chapters, and there's a chance that some crucial point might be missed if a reader simply goes right to a later chapter. Also, a word to the wise: **work out the problems**. Don't just look at an exercise and say to yourself, "I know how to do that." Even worse: do not look at a problem, check the answer, and say to yourself, "I could have gotten that answer." If you do not actually work out the problem and see how to get to the correct answer, you are not learning anything. You learn best by doing. Do. Work out the problems yourself.

The book is (hopefully) straightforward, loaded with examples and exercises (and answers), and easy to read. There is space in the book for you to work out problems on the page, should you so choose. There are also some sections and problems that are set aside specifically for the use of calculators, because knowing how to use your calculator is an important part of understanding the math you use.

The ordering of the chapters mimics the order how some of the material is utilized during the course of most introductory chemistry texts. However, you may need to cover the first four or five chapters quickly, since concepts like balancing equations and the numerical problems that accompany balanced equations require a mastery of all of Chapters 1 - 5. The final chapter on graphing will be particularly useful to you if you are also taking a chemistry laboratory course, since it is common to perform an experiment that requires you to graph your results. While this chapter stands out from the rest as being less computationally oriented, it will also help you in your lecture courses to better understand graphs in your texts.

When I was in school, I had a math teacher say to me, as I recall, "I don't care if you don't like math; you'll be doing it the rest of your life." (Or something along those lines.) Well, she was right. And as useful as math is, it is even more important that it be done correctly.

Good luck, and good chemistry (and math) to you all!

Cleveland, Ohio
January 26, 1996

Chapter 1. Numbers, Units, and Scientific Notation

Introduction

All of science is very mathematical, and math uses numbers. But science does not use numbers alone; it uses units, too.

A <u>number</u> tells you <u>how</u> <u>much</u>; a <u>unit</u> tells you how much <u>of</u> <u>what</u>. In science, both numbers and units are very important. For example, when one is given the quantity "2 grams of iron", then the number is "2" and the unit is "grams of iron":

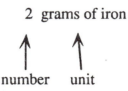

2 grams of iron

number   unit

This may seem like a very trivial distinction, but later we will find that it is a very important one.

It is critical in chemistry to keep track not just of the numbers in a mathematical problem, but also the units. For example, when you work a mole-mole type of problem in chemistry, a frequent question you might have is "Moles of WHAT?" That's because the amount, moles, can refer to anything. Consider, for example, a dozen eggs, a dozen doughnuts, a dozen people, a dozen water molecules. If you were asked how many are present and you answer "A dozen" (meaning twelve), then your answer is incomplete. A dozen OF WHAT? Of course, the question is incomplete; you should be asked how many eggs you have, or how many people are present. In that case, the unit is understood. It's eggs, or people. But too often in chemistry problems, it's easy to lose track of what the exact unit is.

Example 1.1. Identify the number and unit in each of the given quantities. (a) A dozen eggs. (b) Three blind mice. (c) 60 miles per hour. (d) 8850 meters above sea level. (e) 365 days.

Solutions. (a) number = dozen (or 12), unit = eggs. (b) number = 3, unit = blind mice. (c) number = 60, unit = miles per hour. (d) number = 8850, unit = meters above sea level. (e) number = 365, unit = days.

1

More on Numbers

Numbers themselves can be greater than zero (<u>positive</u> numbers) or less than zero (<u>negative</u> numbers). Negative numbers have an explicit negative sign preceding them, like -4 or -200, whereas positive numbers may or may not have a positive sign. For example, the numbers 75 and +75 both refer to <u>positive</u> 75. **In the absence of any sign, always assume that the number is positive.** Zero itself is considered neither positive nor negative.

The numbers that we use to count things (1, 2, 3, 4, 5, ...) are called <u>whole</u> <u>numbers</u> or <u>integers</u>. Integers include 0 as well as negative numbers. The numbering system that we use allows us to combine single digits to make larger numbers to indicate how many 10's, how many 100's, how many 1000's, etc., a number represents. For example, the number 1234 implies one 1000's, two 100's, three 10's, and four 1's. The position of each digit indicates what it represents. We say that the 1 is in the thousands' place, the 2 is in the hundreds' place, the 3 is in the tens' place, and the 4 is in the ones' place:

$$
\begin{array}{c}
\text{\# of 100's} \quad \text{\# of 1's} \\
\downarrow \qquad \downarrow \\
1\ 2\ 3\ 4 \quad = (1 \times 1000) + (2 \times 100) + (3 \times 10) + (4 \times 1) = 1234 \\
\uparrow \qquad \uparrow \\
\text{\# of 1000's} \quad \text{\# of 10's}
\end{array}
$$

Objects that come in packages (like people, eggs, donuts, etc.) are easily counted using integers. However, there are many objects that can be divided into smaller parts. We use <u>decimal</u> <u>numbers</u> to indicate those smaller parts. We use a period, called a <u>decimal</u> <u>point</u>, to separate the integer numbers from the decimal numbers. (In Europe and elsewhere, they use a comma to separate integers from decimal numbers.) For example, if we have two grams plus one half of a gram of iron, we use the decimal number 2.5 to describe the number of grams of iron, where the ".5" is used to represent one half, or five tenths, of a gram. Position again determines what fraction a digit represents. If the position of a decimal number is next to the decimal point, it represents the number of $\frac{1}{10}$ths of a unit. If the position of a decimal number is the second position after the decimal, it represents the number of $\frac{1}{100}$ths of a unit, and so forth. Like above, then, we have for decimal numbers:

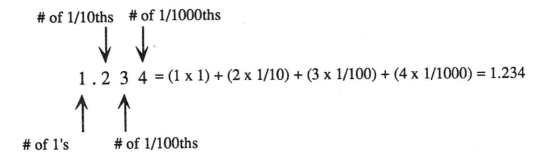

For either integer or decimal numbers, there may be a zero in any position to indicate that there is no contribution from that place. For example, the number 101.01 means "one hundred and one, and one hundredth." However, there are no tens and no tenths in this number. Because our numbering system uses the position of the number as an important way to communicate its value, we have to use zeros to properly position non-zero digits.

Example 1.2. Write the numbers from the following descriptions. (a) Three thousands, eight hundreds, 3 tens, and 9 ones. (b) Five ones, 7 tenths, and 2 hundredths. (c) Negative three ones, 7 hundredths, and 2 thousandths.

Solutions. (a) 3839 (b) 5.72 (c) -3.072. Notice in the third answer the "0" in the tenths' place to properly place the hundredths' and thousandths' places.

More on Units

As mentioned above, the unit attached to a number tells you what specific thing is being quantified. In chemistry, it is very important to keep track of the unit that goes along with the number.

Many people in the United States are very familiar with a particular set of units called English units: pounds, ounces, gallons, feet, miles, degrees Fahrenheit. While there is nothing wrong with these units, it has to be admitted that there is no consistency between them. Scientists all over the world have been using a different set of units. Most nations also have adopted them; the United States is one of the last holdovers using the English units.

Although the other set of units is sometimes called the metric system, the units are more properly called SI units (the SI comes from the French phrase "Le Système International d'Unités"). The SI units are divided into two groups: fundamental units and derived units. There are only seven fundamental units, five of which are commonly used in chemistry:

Fundamental Units

Quantity	Unit	Abbreviation
Mass	kilogram	kg
Amount	moles	mol
Time	second	s (but sec is common)
Temperature	Kelvin	K
Length	meter	m

(The other fundamental units are candela, for luminous intensity, and ampere, for electric current.) The fundamental units also have abbreviations, given in the table, in order to communicate them more efficiently. The kilogram is itself based on the unit gram, which has the abbreviation "g."

Derived units are combinations of the fundamental units. For example, to determine a volume of an object, you multiply its length times its width times its height. To calculate the volume, you not only multiply the numbers of the lengths you measure, but **you multiply the units as well.** If you express your units in meters, then the units of the volume become meter x meter x meter, or meter^3, or more simply m^3. The unit for volume is a derived unit.

Another example is the unit for velocity. Velocity is defined as "distance traveled divided by time taken." Therefore, in terms of fundamental units, velocity has the derived unit of $\dfrac{\text{meters}}{\text{second}}$, or $\dfrac{\text{m}}{\text{s}}$.

Some derived units are more complicated combinations (usually multiplications and divisions) of fundamental units, and it is common to define a new unit to stand for the more complicated combination. (You should never forget, however, that the "new" unit is actually a combination of the fundamental units.) The unit of force is called a newton (abbreviated N) but is a combination of kilogram, meter, and second units:

$$\text{newton} = \frac{\text{kilogram} \cdot \text{meter}}{\text{second}^2}$$

$$N = \frac{\text{kg} \cdot \text{m}}{\text{s}^2}$$

The following table lists other, common derived units and their "new" definitions. You will probably encounter most of them in your study of chemistry.

Redefined Units

Quantity	Fundamental Units	"New" Unit	Abbreviation
Force	$\dfrac{kg \cdot m}{s^2}$	newton	N
Work, heat, energy	$\dfrac{kg \cdot m^2}{s^2} = N \cdot m$	joule	J
Power	$\dfrac{kg \cdot m^2}{s^3} = \dfrac{J}{s}$	watt	W
Temperature	K - 273.15	degree Celsius	°C

The last new unit, for temperature, is based on the fact that the absolute temperature scale, which uses the Kelvin temperature unit, is the same size as the Celsius (or Centigrade) degree, only shifted by a little over 273 degrees. Scientists commonly use the Celsius scale to express temperatures. At the same time, they recognize that many mathematical equations used in chemistry require that temperature be expressed in the Kelvin temperature scale. (But it's simple to change it.) Most nations (and again, the US is almost the only exception) use degrees Celsius to express temperatures as part of their weather report!

Although the SI unit for volume should be m^3, a meter cubed (or "a cubic meter") is a rather large volume. The SI system therefore defines a more manageable unit, the liter, abbreviated L, which is one thousandth of a cubic meter, or $0.001\ m^3$. Chemistry thus uses the liter (a volume that is a little larger than a US quart) as the "basic" unit of volume, even though it is ultimately considered a derived unit.

Having "manageable" units is a primary concern of the SI units. However, despite the definition of the liter, we can still easily imagine circumstances where this volume unit is too large or too small to effectively describe the volume. The same is true of the other fundamental and derived units. However, the SI system also has a series of prefixes that are used to indicate a scaling factor. The easy part of these prefixes is that they all involve either multiplication or division by some power of 10.

The common prefixes and their scaling factors are:

Scaling Prefixes for SI Units

Prefix	Scaling Factor	Abbreviation
giga-	1,000,000,000 x	G
mega-	1,000,000 x	M
kilo-	1,000 x	k
deci-	$\frac{1}{10}$ x	d
centi-	$\frac{1}{100}$ x	c
milli-	$\frac{1}{1000}$ x	m
micro-	$\frac{1}{1,000,000}$ x	μ
nano-	$\frac{1}{1,000,000,000}$ x	n

There are others, but these are the most common ones that you will probably encounter in chemistry.

The prefixes are combined with the SI units to define "new" units that have more manageable numbers. For example, the fundamental unit of mass, the kilogram, has one of the prefixes: "kilo-". The unit "kilogram" means "1000 × grams", or 1000 grams:

$$kilogram$$

$$1000 \times \quad gram \quad = 1000 \text{ grams}$$

The prefixes also have abbreviations, and they are combined with the abbreviations of the units to create a simple abbreviation for the scaled unit. Therefore, the abbreviation for kilogram is "kg" (as indicated above), and it is simply the combination of the abbreviations for kilo- (k) and gram (g).

Example 1.3. State the meanings and give the abbreviations of the following units. (a) kilometer (b) microliter (c) megajoule (d) nanonewton (e) centimeter.

Solutions. (a) A kilometer is 1000 x meter or 1000 meters and has the abbreviation km. (b) A microliter is $\frac{1}{1,000,000}$th of a liter and has the abbreviation μL. (c) A megajoule is 1,000,000 joules and has the abbreviation MJ. (d) A nanonewton is $\frac{1}{1,000,000,000}$th of a newton and is abbreviated nN. (e) A centimeter is $\frac{1}{100}$th of a meter and is abbreviated cm.

Converting from one unit to the other is as simple as moving the decimal point and adding zeros to place the digits in the proper columns. For example, to convert from km to m, the decimal place is moved three places over:

1.54 km into m:

$$1.54 = 1540. \text{ m}$$

where the final zero was added to make sure the decimal point is in the right place. 1.54 km is the same as 1540 m. This makes sense: meters is the smaller unit, so there are more of them in the same distance.

Moving the decimal point in the other direction:

1.54 mg into g:

$$1.54 = 0.00154 \text{ g}$$

where again, zeros have been added to make sure the digits are in the proper column. Notice that the decimal point is moved over one position for every zero in the scaling factor. There are three zeros in "1000 x" and "$\frac{1}{1000}$th", so in both cases the decimal point is moved over three positions. If the scaling factor is a number greater than 1 (as 1000 is in the first example), then the decimal point moves to the right. If the scaling factor is less than one (as $\frac{1}{1000}$ is in the second example), then the decimal point is moved to the left.

Example 1.4. Change the following quantities into the requested units. (a) 0.154 km into m. (b) 8450 grams into kg. (c) 0.000 65 L into mL. (d) 0.000 65 L into μL. (e) 65,000,000 J into MJ. (f) 24.9 GW into W. (g) 0.000 000 03 g into μg.

Solutions. (a) 0.154 km is 154 m. (b) 8450 grams is 8.45 kg. (c) 0.000 65 L is 0.65 mL. (d) 0.000 65 L is 650 μL. (e) 65,000,000 J is 65 MJ. (f) 24.9 GW is 24,900,000,000 W. (g) 0.000 000 03 g is 0.03 μg.

You should become very familiar with the units, the prefixes, and how the prefixes are combined with the units. You should also become familiar with the abbreviations that are used and how they are combined, because it is very common in chemistry to use the prefixes and their abbreviations when using units.

Scientific Notation

The last two conversions in the previous example show that some numbers can get very, very large, and some numbers can get very, very small. (When we say "very small," we mean getting closer and closer to zero. We do not mean negative numbers.) It becomes troublesome to write all of the zeros necessary just to position one or two digits in the correct column. Since chemistry occasionally deals with very large or very small numbers, an easier way to express such numbers is needed.

Scientific notation is a simpler way of writing very large or very small numbers. It is based on powers of 10:

$$1,000,000,000 = 10 \times 10 \times 10 \times 10 \times 10 \times 10 \times 10 \times 10 \times 10 = 10^9$$
$$100,000,000 = 10 \times 10 \times 10 \times 10 \times 10 \times 10 \times 10 \times 10 = 10^8$$
$$10,000,000 = 10 \times 10 \times 10 \times 10 \times 10 \times 10 \times 10 = 10^7$$
$$1,000,000 = 10 \times 10 \times 10 \times 10 \times 10 \times 10 = 10^6$$
$$100,000 = 10 \times 10 \times 10 \times 10 \times 10 = 10^5$$
$$10,000 = 10 \times 10 \times 10 \times 10 = 10^4$$
$$1,000 = 10 \times 10 \times 10 = 10^3$$
$$100 = 10 \times 10 = 10^2$$
$$10 = 10 = 10^1$$

Large numbers are therefore positive powers of 10. The power or exponent is written as a small superscript on the right-hand side of the 10. The power in the numbers above is equal to the number of zeros following the digit 1. Ten raised to the 0 power is, by definition, 1. (Anything raised to the 0 power is 1.) Numbers smaller than one can be described by using negative powers of ten:

$$1 = 10^0$$
$$0.1 = \frac{1}{10} = 10^{-1}$$
$$0.01 = \frac{1}{10 \times 10} = 10^{-2}$$
$$0.001 = \frac{1}{10 \times 10 \times 10} = 10^{-3}$$
$$0.000\,1 = \frac{1}{10 \times 10 \times 10 \times 10} = 10^{-4}$$
$$0.000\,01 = \frac{1}{10 \times 10 \times 10 \times 10 \times 10} = 10^{-5}$$
$$0.000\,001 = \frac{1}{10 \times 10 \times 10 \times 10 \times 10 \times 10} = 10^{-6}$$
$$0.000\,000\,1 = \frac{1}{10 \times 10 \times 10 \times 10 \times 10 \times 10 \times 10} = 10^{-7}$$
$$0.000\,000\,01 = \frac{1}{10 \times 10 \times 10 \times 10 \times 10 \times 10 \times 10 \times 10} = 10^{-8}$$

The spaces between the zeros helps keep track of them more easily, like the commas do for the very large numbers. In this case, the negative exponent equals the number of zeros in the original number, **including the zero to the left the decimal point.**

Very large or very small numbers can be written using scientific notation. For example, 65,000,000 can be written as 6.5 x 10,000,000, which is 6.5×10^7 in scientific notation. We also see that 0.000 000 052 can be written as 5.2 x 0.000 000 01, which is 5.2×10^{-8}.

In writing down a number in scientific notation, the convention is to have a non-zero digit first, followed by a decimal point, then the rest of the non-zero digits, stopping with the last non-zero digit. Then, add the "times" sign, and write down 10 raised to the appropriate power. So, for example, while writing 59,220,000,000 in scientific notation as

$$59.22 \times 10^9$$

is technically not *in*correct, but it does not follow the accepted convention of a single non-zero digit before the decimal point. It is more conventional to write it as

$$5.922 \times 10^{10}$$

in proper scientific notation. By the same token,

$$0.5922 \times 10^{11}$$

while referring to the same number, is also not written according to convention. While it might seem that such conventions are nitpicking, understand that when we all follow the same conventions, it is much easier to communicate ideas and facts. A demand to follow a convention is not unique to chemistry, or even science. All fields have their own particular conventions that allow experts in that field to communicate with other experts more efficiently.

Example 1.5. Express the following numbers in appropriate scientific notation. (a) 66,900,000 (b) 0.005 83 (c) 12,001 (d) -0.000 082 07 (e) -3,141,000

Solutions. (a) 66,900,000 can be written as 6.69 x 10,000,000. Therefore, in scientific notation, the number is written 6.69×10^7 (b) 0.005 83 can be written as 5.83 x 0.001. Therefore, in scientific notation, this number is written as 5.83×10^{-3} (c) 1.2001×10^4 (d) -8.207×10^{-5} (e) -3.141×10^6.

Notice in the last two how the negative sign remains with the number itself, and does not affect the sign of the power.

Example 1.6. Change the units on the following quantities to the requested units, and express the final answer in proper scientific notation. (a) 0.000 45 g into µg. (b) 39.6 kJ into J. (c) 52 m into nm.

Solutions. (a) In going from grams to micrograms, the decimal point is moved six places to the right, so the answer in micrograms is 450 µg. In scientific notation, this would be 4.5×10^2 µg. (b) A kilojoule is 1000 joules, so 39.6 kJ is 39,600 J. In scientific notation, that would be 3.96×10^4 J. (c) There are 1,000,000,000 nm, or 10^9 nm, in a meter, so for 52 meters there are 52×10^9 nm. However, for proper scientific notation, this needs to be changed to 5.2×10^{10} nm. We got this answer by recalling that $52 = 5.2 \times 10^1$, and $10^1 \times 10^9 = 10^{10}$.

Scientific Notation and Calculators

These days, proper use of a calculator is practically mandatory in a chemistry class. "Proper" use, however, demands that you actually <u>know</u> how to work the calculator! You might be surprised to know how many beginning chemistry students **think** they know how to use their calculator properly but actually don't. This section, and several others in the chapters that follow, are devoted to proper use of calculators with respect to some of the topics discussed above. In particular, we need to review how numbers expressed in scientific notation are entered into a calculator.

Because different models of calculators are different, this discussion might not be applicable to ALL calculators. However, it will probably be applicable to almost all calculators. You should have a manual for your calculator, which you should consult for specific directions. However, even if your calculator doesn't work exactly like the ones discussed here, you can probably use these directions and, with your own manual, figure out how they are applicable to your own calculator.

The key to understanding how to enter a number using scientific notation is that **your calculator understands powers of 10**. What you enter into the calculator are the decimal part of the scientific notation expression, called the <u>mantissa</u>, and the power on the 10, which we have called the <u>exponent</u>:

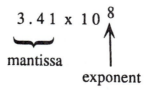

Calculators understand the "times 10 to the" part. All you need to do is enter into the calculator the 3.41 and the 8 correctly. The calculator then understands that you have entered in a number equal to 341,000,000.

When you enter a number into a calculator, it assumes you are entering a decimal-type number – unless you indicate otherwise by pushing a special key. Most calculators have a special key to indicate that the next numbers are the exponent. Look on your calculator for a key that looks like **EE** or **EXP** or, occasionally, **10^x**. You might have to invoke a second-function key to access it. This is the exponent function. When you press it, the next numbers that you enter are the value of the exponent. If you hit the **+/−** key, you negate the <u>exponent</u>. Most calculators only take two or three numbers into the exponent part of the number, along with a minus sign. Also, most calculators do not explicitly show a positive sign if the exponent is positive. Finally, many calculators do not show the "x 10" part of the scientific notation.

For example, suppose we want to enter 3.41 x 10^8 into a calculator. First, press the **3**, the decimal point **.**, the **4**, and the **1** keys on your keypad. The display of your calculator should look something like this:

$$\boxed{\qquad 3.4\ 1 \qquad}$$

Now, press the **EE** or **EXP** or **10^x** key to activate the exponent field. Your calculator display might look like something like this:

$$\boxed{\qquad 3.4\ 1 \quad ^{00} \qquad}$$

where the two zeros will represent the exponent. (Again, keep in mind that your particular calculator might have a different display. You should check your owner's manual if you are uncertain.) **Before touching any other calculator key, enter the exponent value.** You do not have to press the **X** key to indicate multiplication, or even enter 10. The calculator understands that it is 3.41 "times ten to the" correct power. After pressing the **8** key, your calculator display should now look like

$$\boxed{\qquad 3.4\ 1 \quad ^{08} \qquad}$$

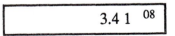

If you need to negate the power – say, the number is 3.41 x 10^{-8} – you need to hit the **+/-** calculator key before you hit any other calculator key that performs an operation (like **X**, **÷**, or even **=**). Hitting the **+/-** key at this point would thus give you

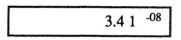

Notice that the negative sign is on the exponent, not the mantissa! You still have a positive number, but it's now equal to 0.000 000 034 1.

At this point, pressing any calculator key that performs an operation – **X**, ÷, or even **=** – brings the calculator function out of the exponent field, and you can continue your calculation. Again, notice that we did not press **X**, nor did we enter 10. The calculator **understands** that the mantissa, 3.41, is being multiplied by 10 raised to a certain power, in this case 8. You, the calculator user, need to understand also how the calculator works. This becomes especially important when you start doing mathematical evaluations using your calculator.

If the number itself – that is, the mantissa – is negative, you have to negate the number **before** you hit your exponent key. Remember, when you activate the exponent portion of the display, any +/- changes the sign on the **exponent**, not the mantissa.

Example 1.7. State the number that is displayed in the following simulated calculator displays, and rewrite them in complete scientific notation and decimal form.

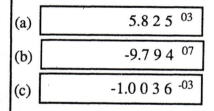

(a)

(b)

(c)

Solutions. (a) This number is 5.825×10^3, which would be written as 5825. (b) This number is a negative number, -9.794×10^7, which could also be written as -97,940,000. (c) This number has a negative exponent: it is -1.0036×10^{-3}, or -0.001 003 6.

Student Exercises

As with any skill, it is important that you (a) understand what you are doing and why, and (b) practice. The following exercises are meant to provide you with practice. Answers are given at the end of the chapter, but in order for you to find out if you truly understand, it is important that you try before seeing the answer! Some space is provided if you want to work in the book directly.

1.1 A patient has a systolic blood pressure of 120 millimeters of mercury. What is the complete unit of this measurement?

1.2. Identify the following units as either fundamental or derived:

(a) centimeter

(b) liter

(c) Kelvin

(d) newton

(e) joule

(f) second

1.3. Change 9.38 km into units of meters, and then change the meters into μm. Express your final answer in scientific notation.

1.4. Change the following quantities into the required units.

(a) 3.09×10^{-3} kW into W.

(b) 9.312×10^5 µs into s.

(c) 6.69×10^{-22} Gm into nm.

(d) 5 L into cL.

(e) 350 K into °C.

(f) 100°C into K.

1.5. Express the following numbers in proper scientific notation.

(a) 93,000,000

(b) 299,790,000

(c) 0.000 000 000 052 9

(d) 602,000,000,000,000,000,000,000

(e) 22.4

(f) 0.000 000 000 000 000 000 000 013 81

1.6. Express the following numbers in decimal form.

(a) 9.65×10^4

(b) 1.09677×10^5

(c) 2.38029×10^2

(d) 2.13×10^5

(e) -8.815×10^{12}

(f) 1.33758×10^{-3}

1.7. Using your calculator, properly enter the following three numbers in scientific notation. When you are done with a number, hit the = key and compare your calculator display to the simulated display in the answer section.

(a) 8.36×10^{12}

(b) 9.104×10^{-25}

(c) -7.772×10^{-9}

Answers to Student Exercises

1.1. The complete unit is "millimeters of mercury".

1.2. (a) derived (b) derived (c) fundamental (d) derived (e) derived (f) fundamental

1.3. 9.38 km = 9380 m = 9,380,000,000 μm = 9.38 x 10^9 μm.

1.4. (a) 3.09 W (b) 0.931 2 s (c) 0.000 669 nm or 6.69 x 10^{-4} nm (d) 500 cL (e) 77°C (f) 373 K.

1.5. (a) 9.3 x 10^7 (b) 2.9979 x 10^8 (c) 5.29 x 10^{-11} (d) 6.02 x 10^{23} (e) 2.24 x 10^1

(f) 1.381 x 10^{-23}.

1.6. (a) 96,500 (b) 109,677 (c) 238.029 (d) 213,000 (e) -8,815,000,000,000 (f) 0.001 337 58

1.7.　(a)

8.3 6 ^{12}

(b)

9.1 0 4 $^{-25}$

(c)

-7.7 7 2 $^{-09}$

Chapter 2.   Arithmetic Evaluation

Introduction

In the course of applying math to chemistry (or any science, for that matter), you will be asked to evaluate mathematical expressions. In many of such tasks, you will have to perform one or more of the four basic arithmetic operations: addition, subtraction, multiplication, and division.

Occasionally, you will have to deal with logarithms and exponentials, but most of the math you will perform will use the four primary operations. Some of these calculations can be done by hand. Many of them will be done using your calculator. Most calculators do not deal with units, so you will have to work those out yourself. This chapter reviews the methods that math has for evaluating various arithmetic expressions.

Addition and Subtraction

<u>Addition</u> is the combination of two quantities. <u>Subtraction</u> is the difference between two quantities. Addition is indicated by the plus sign (+), while subtraction is indicated by the minus sign (-). Addition and subtraction are considered mathematical opposites of each other. Subtraction can be thought of as the addition of a negative number:

$$100 - 23$$
$$= 100 + (-23)$$
$$= 77$$

One key to performing successful addition and subtraction in chemistry is to understand that **all of the quantities being added or subtracted must have the same unit.** Not only must they be the same type of the unit, but they must be the exact same unit. For example, it is obvious that one cannot add 2 kilometers and 5 grams. But some people add 2 kilometers to 5 meters and get 7 . . . kilometers? meters?

In order to add or subtract two quantities, their units must be the same. This may require that you change some of the units:

$$5 \text{ kilometers} = 5000 \text{ meters}$$
$$+ \quad\quad 7 \text{ meters}$$

$$\left. \right\} \text{same unit}$$

$$\overline{\text{TOTAL} = 5007 \text{ meters}}$$

If you wanted to, you could change the 7 meters to kilometers instead:

$$7 \text{ meters} = 0.007 \text{ kilometers}$$
$$+ \ 5 \text{ kilometers}$$

$$\left. \right\} \text{same unit}$$

$$\overline{\text{TOTAL} = 5.007 \text{ kilometers}}$$

and now the sum of the two numbers has units of kilometers. Which answer is correct? THEY BOTH ARE. Since 5.007 kilometers is equal to 5007 meters, both answers are correct. They simply have different units. Unless a problem specifically asks for a particular unit in the final answer, you should be able to choose which appropriate unit you want to determine the answer.

The above example applies to subtractions, too.

Example 2.1. Evaluate the following expression using two different units. (a) 6.8 grams + 0.8215 kilograms. (b) 310 K - 15°C. (For simplicity, we will use the approximate conversion K = °C + 273.)

Solutions. (a) If we want to evaluate the sum in units of grams, we will have to change 0.8215 kilograms into grams. Moving the decimal point over three places, it becomes 821.5 grams. Now we can perform the sum:

$$6.8 \text{ grams} + 821.5 \text{ grams} = 828.3 \text{ grams}$$

If we want to perform the sum using units of kilograms, we have to change 6.8 grams into kilograms. Moving the decimal point over three places, it becomes 0.0068 kilograms. Adding the two quantities:

$$0.0068 \text{ kilograms} + 0.8215 \text{ kilograms} = 0.8283 \text{ kilograms}.$$

You should satisfy yourself that the two answers are in fact the same, only with different units.

(b) We can express a final answer for the difference in the two temperatures in units of Kelvin or degrees Celsius. To convert 15°C into Kelvin,

$$15 + 273 = 288 \text{ K}$$

so that the difference becomes

$$310 \text{ K} - 288 \text{ K} = 22 \text{ K}$$

To express the answer in the other unit, we need to change 310 K to degrees Celsius:

$$310 - 273 = 37°\text{C}$$

Therefore,

$$37°\text{C} - 15°\text{C} = 22°\text{C}$$

The difference has the same value in both Kelvin and degrees Celsius units. That's because the degree size is the same for both scales. (This problem does not imply that 22°C of temperature equals 22 K of temperature! It means that a 22°C <u>change</u> in temperature is the same as a 22 K <u>change</u> in temperature.)

One of the most common subtractions you will perform in chemistry is to calculate the change in some measurement. Part (b) in the above example is a problem of this type. To determine the change in some measurement, **you always take the final value and subtract from it the initial value:**

CHANGE = FINAL VALUE - INITIAL VALUE

Sometimes the symbol Δ is used to indicate the change in some measurement. For example, ΔT would stand for the change in temperature, which is defined as the final temperature minus the initial temperature. Δmass represents the change in mass, which is the final mass minus the initial mass. Changes may have positive or negative values. If the change is positive, it means that the measurement is <u>increasing</u> in value. If the change is negative, then the measurement is <u>decreasing</u> in value.

Example 2.2. Determine (a) the change in temperature if an object goes from 50°C to 12°C, and (b) the change in density from 1.28 g/L to 3.72 g/L.
Solutions. (a) ΔT is equal to the final temperature, 12°C, minus the initial temperature, 50°C. Both measurements have the same units, so we can subtract directly:

$$\Delta T = 12°C - 50°C$$
$$\Delta T = -38°C$$

Notice that the change is negative. This means that the temperature is decreasing.

(b) Δdensity is equal to the final density, 3.72 g/L, minus the initial density, 1.28 g/L. Again, the densities have the same units, so we can evaluate the subtraction directly:

$$\Delta\text{density} = 3.72 \text{ g/L} - 1.28 \text{ g/L}$$
$$\Delta\text{density} = 2.44 \text{ g/L}$$

The change in density is positive, so in this case the density is increasing.

If you performed the above subtractions on your calculator, you should have gotten a negative sign in (a), telling you that the change was negative. Do not forget to include the negative sign as part of your answer!

Multiplication and Division: Numbers

By far, the majority of calculations you will perform in chemistry will involve multiplication and division. You might even perform both in the same calculation.

Recall that multiplication is a short-hand form of addition. Instead of adding 10 to itself four times

$$10 + 10 + 10 + 10 + 10 = 50$$

we say that we are taking 10 five times, and get 50. We write it

$$10 \times 5 = 50$$

which, reading aloud, says "ten times five equals fifty."

Division is the inverse of multiplication. Instead of taking a number many times, you are splitting it into smaller pieces. For example, if we want to separate 50 into five equal parts to get ten, we would write it

$$50 \div 5 = 10$$

which reads "fifty divided by five equals 10."

Instead of using the times sign, x, the act of multiplication is usually indicated differently in some expressions. Sometimes, a simple dot, "·" is used to signify multiplication. For example, writing 7 · 33 implies the multiplication of 7 and 33. Even simpler, if numbers or variables are simply written next to each other, sometimes in parentheses, **multiplication is implied.** Therefore, (7)(-66) implies the multiplication of 7 and -66, and the expression "nRT" implies the multiplication of the variables n, R, and T.

Division is also written differently. Division is commonly represented with a <u>fraction</u>, which simply writes one number above another number, separated by a horizontal line. The number on the top, the <u>numerator</u>, is being divided by the number on the bottom, the <u>denominator</u>. For example, the division of 50 by 5 is written as

$$\frac{50}{5} = 10$$

which is still read as "fifty divided by five equals 10." In this case, 50 is the numerator and 5 is the denominator.

A useful trick (one which we will use in Chapter 4) is to recognize that any number can be thought of as a numerator **divided by one.** For example, the number 10 can be thought of as $\frac{10}{1}$.

With that in mind, we define the <u>reciprocal</u> of a number as the number made by switching the numerator and denominator of any fraction. The reciprocal of $\frac{2}{3}$ is $\frac{3}{2}$:

$$\frac{2}{3} \enspace\curvearrowright\enspace = \enspace \frac{3}{2}$$

Because any number can be thought of as being a fraction with one in the denominator, any number has a reciprocal. The reciprocal of 10 is $\frac{1}{10}$:

$$10 = \frac{10}{1} \enspace\curvearrowright\qquad \text{reciprocal} \enspace = \enspace \frac{1}{10}$$

Multiplication and division can be considered reciprocal operations of each other. Multiplying and dividing a quantity by the same number leaves the original quantity unchanged. For example,

$$3.5 \times 6.7 \div 6.7 = 3.5$$

Another way of writing this is by using a fraction:

$$\frac{3.5 \times 6.7}{6.7} = 3.5$$

The 6.7 in the numerator and denominator cancel each other.

We will occasionally take advantage of the fact that **multiplication by a number is the same as division by the reciprocal of that number. Division by a number is the same as multiplication by the reciprocal of that number.** Sometimes we can simplify expressions by rewriting them with reciprocals. For example:

$$66 \times \frac{1}{4} = \underbrace{66 \div 4}_{} = \frac{66}{4} = 16.5$$

Easier to plug
into calculator?

In this expression, it may be easier for some students to see how to plug numbers into a calculator to evaluate the multiplication operation as a division by the reciprocal. The same tactic can be used when there are fractions within numerators or denominators:

$$\frac{6/7}{3/4} = \frac{6}{7} \div \frac{3}{4} = \frac{6}{7} \times \frac{4}{3} = \frac{6 \cdot 4}{7 \cdot 3} = \frac{24}{21} = \frac{8}{7}$$

where in the center step we are using the fact that multiplication with the reciprocal is equivalent to division.

Example 2.3. Evaluate the following expressions. (a) $6.32 \div \dfrac{8}{14}$ (b) $9 \cdot 8 \cdot 7 \div 5$.

Solutions. (a) $6.32 \div \dfrac{8}{14}$ is the same thing as $6.32 \times \dfrac{14}{8}$, which is 11.06. (b) We can rewrite the expression so that it looks like a fraction: $9 \cdot 8 \cdot 7 \div 5 = \dfrac{9 \cdot 8 \cdot 7}{5}$, which equals 100.8.

One of the nice things about multiplication and division is that they are <u>commutative</u>. This means that if you have an expression where you have to multiply and divide numbers, the exact order in which you perform the functions does not matter. Consider Example 2.3 (b), above. There are several ways we can do the multiplications and divisions:

$$\frac{(9 \cdot 8 \cdot 7)}{5} = \frac{504}{5} = 100.8$$

$$\frac{9}{5} \cdot 8 \cdot 7 = 1.8 \cdot 7 \cdot 8 = 1.8 \cdot 56 = 100.8$$

$$9 \cdot \frac{8}{5} \cdot 7 = 9 \cdot 1.6 \cdot 8 = 100.8$$

and many other possibilities. However, all of these combinations give the same, correct answer! Therefore, when evaluating expressions with many numbers in the numerator and denominator, the order of operation does not matter, as long as only multiplication and division are being performed. We will consider order of operations in more detail later in this chapter.

This is true, of course, only as long as you evaluate the expression properly! With many calculators, there can be a potential problem. Consider the following combination of multiplication and division:

$$\frac{4.6 \cdot 63.9}{2.55 \cdot 34.0}$$

Try plugging this into your calculator. If you got 3.3903...., then you evaluated the expression properly. However, some people will get 3919.199......, which is incorrect. What happened?

For people who get the incorrect answer, this is the usual pattern of keystrokes on their calculator:

$$\textbf{4.6} \quad \textbf{X} \quad \textbf{63.9} \quad \div \quad \textbf{2.55} \quad \textbf{X} \quad \textbf{34.0} \quad \textbf{=}$$

because, of course, that's what the expression is. But what many people don't realize is that on most calculators, every time you hit an operation key (i.e. **X**, ÷, =, etc.) the calculator evaluates the entire expression up to that point! Therefore, when that second **X** key was entered, the calculator evaluated the entire expression up to that point, which was

$$\frac{4.6 \cdot 63.9}{2.55}, \text{ which} = 115.2705....$$

Then, when the second **X** key was hit, the calculator assumes that this number, 115.2705...., was being multiplied by 34.0, **which the calculator assumes is in the numerator!**

Because many calculators evaluate expressions whenever an operation button is pressed, the proper way to evaluate the expression would be to either use your parentheses keys:

4.6   X   63.9   ÷   (2.55   X   34.0   )   =

Or, you can use the ÷ operation instead of the **X** operation for the final number:

4.6   X   63.9   ÷   2.55   ÷   34.0   =

Either way should give you the proper answer.

Some more advanced calculators will not evaluate an expression until the = key is pressed. Usually, these calculators have large viewscreens that show all of the numbers and operations so you can check your expression before hitting =. Many of these calculators also have alphanumeric capability; i.e. you can enter words as well as numbers. If your calculator is like this, much of the above discussion may not apply strictly. You should read your calculator manual to determine what the correct order of entry is to evaluate expressions like the ones above. That's the basic rule for calculator use: **know how to use your calculator!**

Multiplication and Division: Units

Like numbers, we can multiply and divide units too. In fact, we have already seen some examples of that in the last chapter, when we discussed derived units. Units of volume are found using (length) x (length) x (length), and so we have units like m^3 and cm^3 for volume. We call them "meters cubed" or "cubic meter" and "cubic centimeter." When different units are multiplying together, as in N·m, we speak it as "newton-meter." As with numbers, the assumption here is that unless otherwise stated, derived units are assumed to be __products__ of the more basic units. Thus, when you hear "kilowatt-hours," it is correct to understand that this unit is "kilowatts __times__ hours."

Units can also be divided. Perhaps one of the most common examples of this in everyday life is the unit for velocity, miles per hour. Mathematically, the word "per" implies a division, so miles per hour can be written in fraction form as

$$\frac{\text{miles}}{\text{hour}}$$

Similarly, if a unit is given as a simple fraction, it can be stated using the word "per" to indicate the division. Thus,

$$\frac{\text{grams}}{\text{mole}}$$

can be stated as "grams per mole."

Derived units can get much more complicated than this, as suggested by the table of derived units given in Chapter 1. An important point about units, however, is that they follow the same algebraic rules of factoring out of numerators and denominators that numbers do. If the exact same unit is in the numerator and denominator, it can be crossed out of both.

If, for example, you have the following combination of units,

$$\frac{\text{km}}{\text{sec}} \cdot \frac{\text{sec}}{\text{hr}}$$

then the rules of fractions state that the product of theses two fractions is the product of the numerator quantities divided by the product of the denominator quantities. Thus, this equals

$$\frac{\text{km} \cdot \text{sec}}{\text{sec} \cdot \text{hr}}$$

But the unit "sec" appears in both numerator and denominator, so it can be canceled out:

$$\frac{\text{km} \cdot \cancel{\text{sec}}}{\cancel{\text{sec}} \cdot \text{hr}} = \frac{\text{km}}{\text{hr}}$$

In Chapter 4, we will use this idea a lot.

Quite a few derived units may not make much sense at first glance. For example, we are relatively familiar with the derived unit $\frac{km}{hr}$, kilometer per hour, as a unit of velocity. Another example would be $\frac{g}{mL}$ as a unit of density. But some derived units may not make as much sense up front. For example, the unit $\frac{kg \cdot m}{sec^2}$ does not have immediate physical significance to most of us. Neither does $\frac{L \cdot atm}{mol \cdot K}$, which is one possible set of units for the ideal gas law constant, a very important quantity in chemistry.

The reason for these seemingly complex units is <u>mathematical necessity</u>. Recall that **units must follow the same rules of algebra as numbers do**. This means that the algebraic combination of units must be performed along with the algebraic combination of numbers. This requires us to take certain, seemingly unusual, mathematical combinations of seemingly unrelated units.

For example, consider Newton's second law: if a force, F, acts on a body, the magnitude of the force is proportional to the acceleration, a, caused by the force as well as to the mass of the body, m. Mathematically, this is written as $F = ma$. Mass has units of kg, and acceleration has units of m/sec/sec, which can be shown to be equal to $\frac{m}{sec^2}$. (You can show this by applying the rules of reciprocals from the previous section to units, not just numbers!) Therefore, when multiplying a mass time an acceleration, you also have to multiply their <u>units</u>: $kg \cdot \frac{m}{sec^2}$, or $\frac{kg \cdot m}{sec^2}$. This means that force must have units of $\frac{kg \cdot m}{sec^2}$. It does, and for simplicity's sake we rename this derived unit the <u>newton</u> and give it the symbol N. Ultimately we use the newton as the unit of force, but we must always remember that it is originally defined as $\frac{kg \cdot m}{sec^2}$.

Not all derived units like this get renamed. For example, $\frac{L \cdot atm}{mol \cdot K}$, which we will use in ideal gas problems, is not given a new name. However, the algebra we use to manipulate these units, as we will see in Chapter 6, requires that certain quantities have such unusual derived units.

Example 2.4. Determine the overall unit for the final answer that you will get by performing the following operations. (a) Mass is multiplied by velocity to give a quantity called momentum. (b) Electric charge, which has units of coulombs, C, is squared and then divided by distance to get energy. (c) Current, which has units of amps, is multiplied by time in seconds to get units of electric charge in units of coulombs, C.

Solutions. (a) With mass having units of kg and velocity having units of $\frac{m}{sec}$, momentum has units of kg$\cdot\frac{m}{sec}$ or $\frac{kg\cdot m}{sec}$. (b) Squaring coulombs gives C^2, and dividing that by m, the unit of distance, gives $\frac{C^2}{m}$ as a unit of energy. [NOTE: This derived unit is not equal to J, the SI unit of energy. However, there is a conversion factor. Check your textbook to see if it deals with these topics.] (c) Current times time gives units of amp·sec, which according to the example is a definition of the unit coulomb: C = amp·sec.

One very important consideration is the fact that **most calculators do not work with units, only with numbers.** It is up to you, the student, to interpret the way the units in a problem are handled. Sometimes this takes a little practice. However, in the long run you will find that a proper handling of the units in these algebraic problems is not only necessary, but can sometimes actually assist you in figuring out how to work a problem. Chapters 4, 5, and 6 will give many examples of how we use units to help work a problem.

Algebra and Equations

A portion of the mathematical skills you will be performing in chemistry is using a given equation to solve for some unknown quantity. These equations are composed of variables whose meanings, at least, you should be familiar with. In all cases, you will probably be asked to solve for one of the variables when all of the other variables will be known to you (or you can find them out).

An <u>equation</u> is any mathematical expression that contains an equals sign, =. We have already seen some very simple forms of equations, like

$$\frac{50}{5} = 10$$

An equation implies that what is on the left side of the = sign is the same as what is on the right side of the = sign. In the above example, we recognize that the fraction $\frac{50}{5}$ reduces to $\frac{10}{1}$ which equals 10, so the overall equation reduces to "10 = 10," which we know to be true. On the other hand, if we had the equation

$$\frac{55}{5} \overset{?}{=} 10$$

then we know that something is wrong because $\frac{55}{5}$ does not equal 10. This is not a proper equation.

Many equations are composed of variables, not numbers. <u>Definitions</u> are among the simplest of equations of variables. For example, the definition of average velocity can be written as:

$$\text{average velocity} = \frac{\text{distance traveled, in m}}{\text{time taken, in sec}}$$

Since the quantity on the left must equal the quantity on the right, if we have distance and time, we can determine the average velocity. If the distance were 400 m and it took 10 seconds to travel that distance, then

$$\text{average velocity} = \frac{400\,m}{10\,sec} = \frac{400}{10}\frac{m}{sec} = 40\frac{m}{sec}$$

where we are showing that $40\frac{m}{sec}$ is equal to $\frac{400\,m}{10\,sec}$.

One of the things that you must be able to do is to solve for <u>any</u> variable in an equation, if you are given all of the others. If your average velocity were 25 meters per second and you needed to travel 5000 meters (which is five kilometers), how many seconds would that take you?

Since the quantities involved are the same as used for the definition of average velocity, **you can use the same equation to determine an answer.** However, you know different quantities – and you are seeking to solve for a different quantity, also. What you have is

$$25\frac{m}{sec} = \frac{5000\,m}{\text{time taken}}$$

and you have to solve for the time, in units of seconds.

There are two keys to being able to solve for any variable in an equation:

- The quantity you are looking for **must be all by itself on one side of the equation** (it doesn't matter which side);
- The quantity you are looking for **must be in the numerator**; that is, you should be able to write the quantity as a value divided by one in the denominator.

Performing mostly multiplication and division, you can rearrange any equation to isolate the desired quantity by itself in the numerator; then simply evaluate the numbers and units on the other side of the equation to get your final answer.

For the question above, the quantity you are looking for (sometimes referred to as "the unknown" in an equation) is "time taken." You need to rewrite the equation algebraically to isolate that quantity by itself in the numerator on one side of the equation. (It's in the denominator right now.) There are several ways to do this algebraically. What follows is not the only way. But, if you perform all of the algebra properly (no matter what way you do it), you should get the same answer.

First, we will multiply both sides of the equation by the variable "time taken." We can do this because, when starting with an equality, if you perform the same operation to both sides of the equality, it is still an equality. (The only exception is multiplying or dividing by zero.) We get

$$25\frac{m}{\text{sec}} \cdot \text{time taken} = \frac{5000\ m}{\text{time taken}} \cdot \text{time taken}$$

Now consider the right side of the equation. Because we have "time taken" in the numerator and the denominator of the fraction, they can cancel out:

$$25\frac{m}{\text{sec}} \cdot \text{time taken} = \frac{5000\ m}{\cancel{\text{time taken}}} \cdot \cancel{\text{time taken}}$$

$$25\frac{m}{\text{sec}} \cdot \text{time taken} = 5000\ m$$

Now, we divide both sides of the equation by $25\ \frac{m}{\text{sec}}$. Again, we can do this to both sides of the equation and still have an equality:

$$\frac{25\frac{m}{sec} \cdot \text{time taken}}{25\frac{m}{sec}} = \frac{5000 \text{ m}}{25\frac{m}{sec}}$$

Notice that we are including the units along with the numbers. On the left side, the $25\frac{m}{sec}$ is in the numerator and denominator, and it cancels out:

$$\frac{25\frac{\cancel{m}}{sec} \cdot \text{time taken}}{25\frac{\cancel{m}}{sec}} = \frac{5000 \text{ m}}{25\frac{m}{sec}}$$

We have canceled the 25 as well as the units $\frac{m}{sec}$! What we have left is

$$\text{time taken} = \frac{5000 \text{ m}}{25\frac{m}{sec}} = \frac{5000}{25}\frac{m}{\frac{m}{sec}}$$

where now we have what we are looking for, the time taken, all by itself on one side of the equation and in the numerator. (Remember, it can be thought of as a fraction with one in the denominator.) Using our calculator, we evaluate the numerical part of the answer on the right-hand side: $\frac{5000}{25} = 200$. To evaluate the units, which most calculators will not do, we recall that a division can also be considered as multiplication by the reciprocal:

$$\frac{m}{\frac{m}{sec}} = m \div \frac{m}{sec} = m \times \frac{sec}{m} = sec$$

where in the last step, the meter unit has been canceled out from the numerator and the denominator. The final unit on the answer is sec, so the complete answer is

$$\text{time taken} = 200 \text{ seconds}$$

One thing to keep in mind with regard to units. **They make sense.** We were looking for an amount of time, and the answer has a unit of time, seconds. If we performed the units analysis (sometimes called dimensional analysis) incorrectly and got meters for our answer, we would say that the time taken was 200 meters. HUH? That doesn't make sense! "Meters" is a unit of length, not a unit of time. **Always ask yourself if the unit you get for an answer is consistent with the quantity you are looking for.** In this way, keeping track of units can actually help you work out a solution to a problem.

Example 2.5. The density of a material is defined as the mass of the material divided by its volume. (a) Write the mathematical equation for the definition of density. (b) What is the density of mercury if 24 milliliters has a mass of 326.4 grams? Express the density in those units. (c) How many milliliters of osmium are needed to have a mass of 100 grams if its density is 22.4 grams per milliliter? (d) What is the mass of 1.1×10^6 L of hydrogen gas if it has a density of 0.0899 grams per liter?

Solutions. (a) Using the definition of average velocity as a guide, we see that the mathematical definition of density can be written as

$$\text{density} = \frac{\text{mass}}{\text{volume}}$$

(b) The density of mercury is found by plugging into the definition of density:

$$\text{density} = \frac{\text{mass}}{\text{volume}} = \frac{326.4 \text{ g}}{24 \text{ mL}} = 13.6 \frac{\text{g}}{\text{mL}}$$

(c) In order to determine the volume of osmium necessary, we set up the density equation as

$$22.4 \frac{\text{g}}{\text{mL}} = \frac{100 \text{ g}}{\text{volume}}$$

The manipulations of this equation are similar to the velocity example from above: multiply both sides by volume, then divide both sides by the value of the density. We get:

$$\text{volume} = \frac{100 \text{ g}}{22.4 \frac{\text{g}}{\text{mL}}} \approx 4.46 \text{ mL}$$

where we use the "≈" sign to mean "approximately equal to," since we have limited our answer to three digits. We check our final unit and recognize that mL is in fact a unit of volume.

(d) The mass of hydrogen for the volume given takes only one rearrangement step:

$$0.0899\frac{g}{L} = \frac{mass}{1.1\times10^6\,L} \quad \Rightarrow \quad 1.1\times10^6\,L \cdot 0.0899\frac{g}{L} = mass$$

Multiplying the two quantities together and canceling out the L unit, we find that the mass = 98,890 grams of hydrogen, or just under 99 kg.

Order of Operations

Addition, subtraction, multiplication, division, exponents and logarithms, etc., are called operations. They ask that you do something to the numbers or expressions on either side of the operation sign (i.e. +, -, X. ÷, log, etc.), which are called operators.

When trying to figure out an expression that has a combination of operators, you cannot simply evaluate them in any order. For example, the following expression contains addition and multiplication (in terms of an exponent):

$$(2 + 3)^2$$

This expression cannot be evaluated properly by distributing the square through the parentheses, squaring the numbers, and adding them together. It is **incorrect** to evaluate this expression this way:

$$(2 + 3)^2 = (2^2 + 3^2) = 4 + 9 = 13 \quad \text{WRONG!}$$

This is incorrect. Exponents are not distributed through an expression in parentheses. In this case, the square requires that we evaluate

$$(2 + 3)^2 = (2 + 3)(2 + 3) = (5)(5) = 25$$

which is the correct way to evaluate the original expression.

The rules of algebra are set up so that a particular order of evaluating operations is required to evaluate a complicated expression correctly. The correct order of operations is

- **Parentheticals** – evaluate the expressions inside parentheses
- **Raised powers** – perform any exponential operations (or inverse exponential operations, like square roots and logs)
- **Multiplication & Division** – perform any products or quotients
- **Addition & Subtraction** – combine numbers or variables together by adding or subtracting

Notice that Multiplication & Division and Addition & Subtraction are grouped together in the same steps. That's because algebra recognizes the inverse relationship between these operations. In order to remember the order Parentheticals, Raised powers, Multiplication & Division, and Addition & Subtraction, some people use the mnemonic "Please Remember My Dear Aunt Sally". The first letter in each word, PRMDAS, stands for the type of operation. Notice that while this mnemonic device puts division after multiplication and subtraction after addition, division and multiplication are actually evaluated together, since they are in reality inverse operations. Addition and subtraction are also actually evaluated together, since they too are really inverse operations.

It is important to realize that in some complicated expressions, you will have an expression inside an expression inside an expression . . . etc. These are called <u>nested</u> expressions. The rule for the order of evaluating nested expressions is to **evaluate the innermost expression first and work your way to the outermost expression in steps.**

It is not often that you will have to deal with all of the algebraic operations in any one problem. It is difficult, then, to come up with relevant examples that illustrate all of the possibilities. Hopefully, the few given here will suffice to give you the general idea.

To evaluate the expression

$$K = \frac{(2.0)^2 (3.0)}{(0.1)^2 (0.5)^3}$$

we first evaluate the raised power for each number inside each set of parentheses. $(2.0)^2 = 4.0$, $(0.1)^2 = 0.01$, and $(0.5)^3 = 0.125$. We now have

$$K = \frac{4.0 \cdot 3.0}{0.01 \cdot 0.125}$$

Now we can evaluate the multiplication and division operations to get

$$K = 9600 = 9.6 \times 10^3$$

as the correct answer. You might want to try plugging this into your calculator to verify that this is indeed the correct numerical answer. Suppose we have a more complicated expression, like

$$E = 0.46 - \frac{(8.314)(298)}{(2)(96,500)} \cdot \ln\left(\frac{(1)(2)^3}{(0.5)^2}\right)$$

There are several parts to this one expression, which need to be worked out before everything can be brought together. (This is why it's nice to have a calculator that has a memory function, which you should know how to use properly.) The "ln" is the natural logarithm, and the expression asks that you take the natural logarithm of an expression. (Logarithms will be covered in more detail in Chapter 7.) This is an example of nested expressions. The following shows the order we will take to evaluate the entire expression:

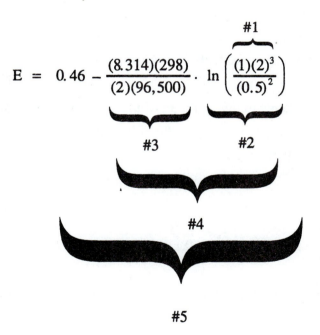

Step #1 evaluates the nested expression inside the logarithm. Evaluating the raised powers and then performing the multiplication and division (which is the correct order of operations), we get

$$\left(\frac{(1)(2)^3}{(0.5)^2} \right) = \left(\frac{1 \cdot 8}{0.125} \right) = 64$$

Step #2 evaluates the natural logarithm of the inside expression, which equals 64. On most calculators, this is performed by entering **64** and then pressing the natural logarithm key, which usually looks like **ln** or **LN**. On a few calculators, like high-powered engineering or graphing calculators, you might have to enter **ln 64** and then press the **=** or **EXECUTE** key. (Check your calculator manual for instructions about how to take logarithms properly using your particular model.) You should get

$$\ln 64 = 4.1588...$$

We will round off to three digits and approximate this as 4.16. Step #3 calls for the evaluation of the fraction in front of the logarithm. If you perform this properly using your calculator, you should get

$$\frac{(8.314)(298)}{(2)(96,500)} = 0.0128...$$

Step #4 has us multiplying this expression by the logarithm term. Notice that this is still a multiplication, so we are on the same step in our PRMDAS scheme.

$$0.0128 \cdot 4.16 = 0.053\ 248$$

and the final step, #5, has us evaluating the subtraction:

$$E = 0.46 - 0.053\ 248 = 0.406\ 752$$

which we will limit to two decimal places to get E = 0.41. (We have not considered units in this example. This was intentional.) Please note that this example followed the conventions of nested expressions and the proper order of operations, PRMDAS, as required. When you develop the right level of sophistication after enough practice, you might be able to evaluate an expression like this using your calculator in one long set of keystrokes. But, don't force it: it is better to evaluate complex expressions in steps, correctly, than to try and do it in one long step – but get the incorrect final answer!

Student Exercises

Space is provided for you to work out the problems here in the book. Answers to all student exercises are given at the end of the exercises, on the next page. But give each problem your best effort before you check yourself!

2.1. Evaluate the following expressions twice, using two different units.

(a) 4.50 kg + 870 g

(b) 923 mg + 2.980 g

(c) 124 cm + 23.00 m

(d) 1.775 L - 43 mL

(e) 350 K - (-43°C)

2.2 (a) Write 45 - 12 as an addition and evaluate.

(b) Write -700 + (-375) as a subtraction and evaluate.

(c) Write 650 + 350 as a subtraction and evaluate.

2.3. (a) What is the change in position of a football player going from the 10-yard line to the 50-yard line?

(b) What is the change in temperature of a cup of coffee going from 75°C to 12 °C?

(c) What is your change in altitude going from +8850 m (above sea level) to -77 m (below sea level)?

(d) What is the change in mass if a plant grows from 70.4 g to 160.3 g?

2.4. Write the following expressions as fractions and evaluate them.

(a) 6.6 x 3.0 ÷ 0.18   (b) (99 x -3.6) ÷ (22.07 x 0.007)

(c) $1 \div (467 + 33)$

(d) $1 + (1 \div (1 + (1 \div (1 + 1))))$

2.5 Write the reciprocal expressions for 2.4 (a) - (c) and evaluate them. Without writing the reciprocal expression, what is the value of the reciprocal of 2.4 (d)?

2.6 (a) Rewrite the expression $6 \div \dfrac{2}{3}$ as a multiplication and evaluate.

(b) Rewrite the expression $\dfrac{22}{7} \times \dfrac{1}{11}$ as a division and evaluate.

2.7. Does 2 x 3 x 4 x 5 equal 5 x 4 x 3 x 2? Why or why not?

2.8. Use your calculator to evaluate the following expressions without writing anything down.

(a) $\dfrac{3}{4 \cdot 5}$ (b) $\dfrac{(9.5)(-44.2)}{50.4}$ (c) $\dfrac{(0.005\ 56)(4567)}{(39.06)(120.6)}$ (d) $\dfrac{(23+95)(54-98)}{(87)(21)}$

2.9. What units should you get when you:

(a) divide units of time by units of velocity?

(b) multiply J·sec by $\dfrac{1}{\sec}$?

(c) divide newtons by $\dfrac{m}{\sec^2}$? (HINT: What are the units that make up the newton unit?)

2.10. How long will it take to travel 50 km at an average velocity of 25 meters per second?

2.11. Evaluate the following expressions. (a) $88 - \dfrac{(2.5 + 7.7)^2}{-3.8}$ (b) $\sqrt{\dfrac{67 - (18 + 33)}{-\left(1 - \left(\dfrac{4}{0.0069}\right)\right)}}$

(c) $(2.9)(44)(-0.072) \div (11.8 - 5)$.

2.12. Evaluate the following expressions.

(a) $\dfrac{(2)(0.08205)(298)}{0.987}$

(b) $100 \cdot \sqrt{\dfrac{1}{1 - \left(\dfrac{2 \times 10^8}{3 \times 10^8}\right)^2}}$

(c) $-0.76 - \dfrac{8.314 \cdot 298}{3 \cdot 96{,}500} \ln \dfrac{(0.75)^2 (0.50)}{(1.75)^3 (1.05)}$

2.13. Is either addition or subtraction commutative? Why or why not?

Answers to Student Exercises

2.1. (a) 5.37 kg, 5370 g (b) 3.903 g, 3903 mg (c) 24.24 m, 2424 cm (d) 1.732 L, 1732 mL (e) 120°C, 120 K.

2.2. (a) 45 + (-12), which equals 33. (b) -700 - (+375) or -700 - 375, which equals -1075. (c) 650 - (-350), which equals 1000.

2.3. (a) 40 yards (b) -63°C (c) -8927 m (d) 89.9 g.

2.4. (a) 110 (b) -2306.9.... (c) 0.002 (d) 1.666 66.....

2.5. The values for the reciprocals are (a) 0.009 090.... (b) -0.000 433 47... (c) 500 (d) 0.6.

2.6. (a) 9 (b) $\frac{2}{7}$.

2.7. Yes, because multiplication is commutative; it doesn't matter what order the values are multiplied in.

2.8. You should get (a) 0.15 (b) -8.3313.... (c) 0.005 390... or 5.390... x 10^{-3} (d) -2.8418...

2.9. (a) You get units of distance, like m. (b) J (c) kg.

2.10. 2000 seconds, or 0.556 hours.

2.11 (a) 115.37..... (b) 0.1662... (c) 1.351....

2.12. (a) 49.54... (b) 134.16.... (c) -0.734...

2.13. Addition is commutative if all of the numbers you are combining are added together, because the order the numbers are added in does not matter. Subtraction, however, is not commutative. As a simple example, consider 2 - 1 and 1 - 2. They do not equal the same number. However, addition of negative numbers is commutative. For example, 2 + (-1) is equal to (-1) + 2.

Chapter 3. Significant Figures

Introduction

Perhaps one of the most unusual concepts that beginning chemistry students need to come to grips with is the idea of significant figures, or "sig figs." Significant figures arise in two situations: in doing mathematical calculations and in making measurements in, say, a chemistry lab. The best way to illustrate the need to follow the rules for assigning sig figs is by example. Suppose you are given that 23 grams of potassium perchlorate, $KClO_4$, has a volume of 9.1 mL. You are asked to find the density. You plug the numbers into the density formula:

$$\text{density} = \frac{\text{mass}}{\text{volume}} = \frac{23 \text{ grams}}{9.1 \text{ mL}}$$

Satisfied that you set up the problem correctly, you use your calculator and evaluate the fraction $\frac{23}{9.1}$ and, reading the numbers from the calculator, get 2.52747252747.... $\frac{g}{mL}$. Now ask yourself: does it make sense to include all of these digits when your original numbers, 23 and 9.1, only had two digits each? No, it does not, and the concept of significant figures addresses that.

Suppose you are asked to make a measurement, like reading the temperature in Celsius from the thermometer below:

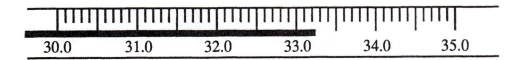

where the thick line represents the mercury in the thermometer. Since there are ten divisions between each number (count them and see), each division represents a tenth of a degree, 0.1°C. Well, you know that the temperature is at least 33.0°, and the mercury is just beyond the line that would represent 0.2, so the temperature is at least 33.2°C. You can guess that the edge is about a third of the way to 0.3, so you might estimate that the edge is reading about 0.23, so the overall temperature is about 33.23°C. How about estimating the thousandths' place in the temperature so it's even more accurate? Well, you can, but the number wouldn't have any real meaning, or <u>significance</u>, because you had to guess at the value of the

41

hundredths' digit. Reporting a temperature to the thousandths' place is pointless, and so we only read the temperature to the hundredths' place.

The concept of significant figures deals with both of these issues. All you have to do is learn the rules and apply them properly. These rules aren't arbitrary. As the above two examples show, the ideas of sig figs is simply a matter of applying common sense when dealing with numbers.

Significant Figures with Measurements

In making measurements in a lab, there are two different ways a number can be presented. It can be given to us in <u>digital</u> fashion, like with numbers in a display. An electronic balance is an example of something that gives numbers digitally. Measurements can also be displayed in <u>analog</u> fashion, which is more pictorial. The thermometer in the above example is one type of analog display of data. Digital displays are usually set to the limits of the measuring device, and we typically use all of the digits they provide. (The concepts of accuracy and precision are important for such devices, but we will not consider those ideas here.) For analog measurements, we have to use our own common sense to determine how many digits we can reasonably report for that measurement.

The rule for significant figures in measurements is simple. The digits that are <u>significant</u> in a measurement are **those decimal places that you know for sure, PLUS the first uncertain decimal place.** Consider the thermometer again:

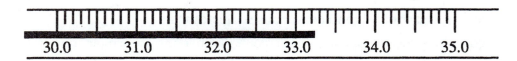

We **know** that the temperature is at least 30°, and we **know** that it is at least 3 degrees above 30. Therefore, both the digits in 33° are significant. We **know** that the temperature is at least higher than 0.2 ° above 33°, so the two in the tenths' place is significant. We are sure, then, that the temperature is at least 33.2°.

We **do not know** the exact position of the mercury to get the digit for the hundredths' place. So, we estimate it at, say, three-tenths of the way between 0.2 and 0.3. We therefore estimate that the final significant figure is 3 in the hundredths' place, and we report the temperature as 33.23°C. We recognize that the final digit might be 2 or 4, but we are fairly certain it is not 8 or 9. But, since we are making a guess about the hundredths' place, any attempt to report a digit for the thousandths' place would be ludicrous. So, we stop at the hundredths' place and report a value that contains 4 significant figures. By the rule of sig figs for measurements, we understand that the final digit is uncertain to (usually) ±1.

Example 3.1. Give proper values for measurements in the following cases.

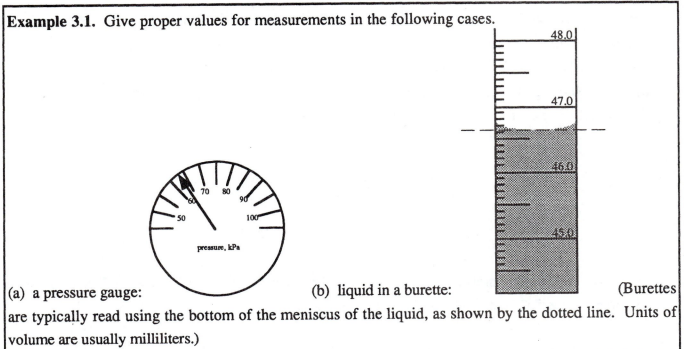

(a) a pressure gauge: (b) liquid in a burette: (Burettes are typically read using the bottom of the meniscus of the liquid, as shown by the dotted line. Units of volume are usually milliliters.)

Solutions. (a) The arrow on the pressure gauge is between the labeled "60" line and the unlabeled "65" line. We know, then, that the pressure is at least 60 but less than 65. We have to estimate the position of the arrow, which seems a little more than halfway to 65, so let us call this 63. We read the pressure as 63 kPa. We only get two significant figures from this measurement. (b) The line marking the bottom of the meniscus of the liquid is above the 46.0 line. The mid-length line represents 46.5, and the line is above that mark. It is also above the short mark indicating 46.6. However, it is not up to the 46.7 mark, so the volume reading will be between 46.6 and 46.7. We estimate the line as being just under halfway, and so judge it as four tenths of the way. Therefore, we have 4 for the hundredths' place. Our final measurement is 46.64 mL, and we have four significant figures in this measurement.

Determining the Number of Significant Figures

Often you will be given a set of numbers, with units, that you will have to use algebraically to determine some final result. Your answer should have only a certain number of significant figures. The number of sig figs in your answer is determined by the number of sig figs in your initial numbers. But how do you determine the number of significant figures in a number?

There are a few simple rules:

- **All non-zero digits are considered significant figures.** The number 345 has three significant figures.

- **All zeros between non-zero digits are significant.** Ignore the decimal point, if necessary. The number 305 also has three significant figures. So does 3.05, because the zero is between non-zero digits.
- **Zeros at the beginning of a decimal number or at the end of a large number are <u>not</u> significant.** For example, 546,000 has only three significant figures. The three zeros simply serve to put the 5 in the 100,000s place, the 4 in the 10,000s place, and the 6 in the 1000s place.

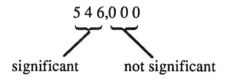

Similarly, the decimal number 0.000528 has only three significant figures, since the four zeros serve only to place the 5, 2, and the 8 in the correct column.

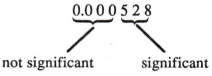

The number 0.0006098 has four significant figures, because we **do** include the zero between the 6 and the 9. However, we do not include the first four zeros as significant.

- **Zeros written at the end of a number <u>after a decimal point</u> are significant.** Otherwise, they would not even be given. For example, 0.67 has two significant figures, but 0.670 has three significant figures. If that final zero were not significant, it should not have been written down.

Zeros cause the biggest problem in determining significant figures: are they, or aren't they? There is a very simple test, though. **Write the number in proper scientific notation.** The number in scientific notation loses the non-significant zeros, and the number still has the same number of significant figures **in the mantissa.** For example, writing 546,000 in scientific notation gives you 5.46×10^5. The "5.46" gives you three sig figs, which is what we determined using the third rule above. Similarly, 0.0006098 can be rewritten as 6.098×10^{-4}, and the "6.098" gives you four significant figures.

Example 3.2. What are the number of significant figures in the following quantities? (a) 2.02 liters (b) 39.36 cm (c) 6.022×10^{23} atoms (d) 2.99790×10^8 m/sec (e) 5,280 feet (f) 0.00150 cm.

Solutions. (a) Because the zero is between two non-zero numbers, it is significant and this quantity has three significant figures. (b) The four non-zero digits in this value means it has four significant figures. (c) The mantissa of this number, written in scientific notation, has four significant figures. (d) The mantissa of this number has six significant figures. The final zero, which is after the decimal point, is significant because of the fourth rule from above. (e) This number only has three significant figures. The final zero serves only to put the non-zero digits in the proper column. Also, you could write this number as 5.28×10^3, which only has three significant figures. (f) 0.00150 has three significant figures. The first three zeros are not significant. However, the last one is (by the fourth rule above), or else it would not even be written.

Some textbooks introduce a few other tricks with regard to zeros being significant. If a number ending in zeros is written with a decimal point, then those zeros on the end are considered significant. For example, 5280 only has three significant figures, but if it is written 5280. with an explicit decimal point, then the zero is considered significant and the number now has 4 sig figs. Some books might also use a line above a zero to indicate that it is the last significant figure in the number. The number 528,000 has three significant figures, but if it's written like 528,0̄00, then the line over the fifth digit indicates that it is the last significant digit, and so this number should be considered as having five significant figures. **You need to determine if your textbook uses either of these conventions.** If you are in doubt, ask your instructor.

Significant Figures in Calculations: Addition and Subtraction

As stated earlier, when you are doing calculations with numbers, the number of significant figures in your answer is going to be dictated by the number of sig figs of the amounts you start with. There are rules for determining the acceptable significant figures in your answers, so all you need to do is know and apply the rules. The rules for addition and subtraction are different than the rules for multiplication and addition, so we will go over those first.

The rule for addition and subtraction is based on the **position of the digits** of the numbers being added or subtracted. The rule is that **the right-most significant position in the answer is that which is common to all of the numbers being added or subtracted.**

It is best illustrated by example. Suppose you are adding 1.00797 and 15.9994 (which are the average atomic weights of hydrogen and oxygen). If you add them using your calculator, you will get the final answer as

$$\begin{array}{r} 1\,5.9\,9\,9\,4 \\ +\,1.0\,0\,7\,9\,7 \\ \hline 1\,7.0\,0\,7\,3\,7 \end{array} = \text{final answer}$$

(Notice how we lined up the decimal points. This will help us determine the correct number of significant figures.) However, the numbers being added together only have four decimal places in common. Therefore, the final significant position is the fourth decimal place, and any numbers after that are discarded:

right-most
common position

$$\begin{array}{r} 1\,5.9\,9\,9\,4 \\ +\,1.0\,0\,7\,9\,7 \\ \hline 1\,7.0\,0\,7\,3\,7 \end{array}$$ » answer limited to 1 7.0 0 7 4
(rounded up)

Now we have limited the final answer to four decimal places and have rounded up the last digit to 4. (We will go over the rules for rounding shortly.)

This rule is very easy to apply even for several numbers, especially if you write all of the numbers with the decimal point aligned. For example,

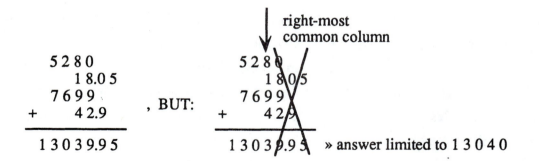

where we have rounded up for the final answer. In this case, our significant figure stops at the tens place and we only have four significant figures in our final answer. (Keep in mind that the zero in 5280 is not significant. That's why we used the tens' column as the right-most significant position.)

There is no significant figure consideration for the left-most column of numbers, only the right-most column. Determination of the proper significant figures for subtractions follows the same rules as addition.

The last step in determining the final answer is <u>rounding</u>. A decision has to be made whether the last significant digit remains the same or if it is rounded up to the next highest number. In both of the above examples, the final significant digit was rounded up. What are the rules for rounding?

- **If the digit <u>after</u> the last significant digit is less than 5, keep the last significant digit the way it is.** This is sometimes called <u>rounding down</u>. For instance, rounding 13,672 to four significant digits gives 13,670, because 2 is less than five.

- **If the digit <u>after</u> the last significant digit is greater than 5, round up the last significant digit to the next highest number,** keeping in mind that rounding up a 9 to a 0 also changes the column to the left. For example, rounding off 1.0058 to four significant figures makes our final answer 1.006. Also, rounding 3.799 to three significant figures brings it to 3.80 (NOT 3.70!).

- **If the digit after the last significant digit is equal to 5,** there are two common rounding methods. (a) Always round up. (b) Round up or down, but to whichever digit is even, not odd. **You should check with your textbook and your instructor to find out which is the preferred method for your course!**

Example 3.3. Evaluate the following combinations to the correct number of significant figures, and correctly round your final answer. (a) 6.7003 - 6.329 (b) The sum of 1.0079, 12.011, and 14.0067 (c) $5.430 \times 10^3 + 4.229 \times 10^2$.

Solutions. (a) The answer for this subtraction is 0.3713, which we have to limit to three decimal places. We therefore round down to 0.371. (b) This sum comes out to be equal to 27.0256, which we have to limit to three decimal places. Because 6 is greater than 5, we round up to get a final answer of 27.026. (c) The way to evaluate this final answer is to write the numbers in non-scientific notation: 5430 + 422.9, which equals 5852.9. However, the way the original numbers are written, the zero in 5430 is significant, so both numbers share significant digits to the units' place. We must limit our answer to having significant digits to that column also. Therefore, the proper answer is 5853.

Significant Figures in Calculations: Multiplication and Division

Perhaps the more common operations you will be performing involve multiplication and division. In this section, we consider how to determine the significant figures for your answers involving those tasks.

First, we need to define <u>exact numbers</u>. Numbers that are used to express quantities that have a known, specific value are called exact numbers and are considered as having an infinite number of significant figures. For example, we know that there are 100 cm in 1 m:

$$100 \text{ cm} = 1 \text{ m}$$

This relationship comes from the definition of the prefix "centi-" and is exact. There are exactly 100 cm in 1 m. We could write it as

$$100.0000000000000..... \text{ cm} = 1.0000000000000.... \text{ m}$$

with each number having any number of zeros after it, but that would be silly. Instead, we simply learn to recognize exact numbers in our calculations and ignore them for the purposes of significant figure determinations. Exact numbers can also have non-integer values. For example, the relationship that 1 inch = 2.54 cm is a defined relationship and so the 2.54 is considered "exact" and is not considered part of the significant figure determination.

A good rule of thumb is that defined quantities and integers are exact, but any property based on a measurement (like mass, volume, pressure, temperature, density) is treated as inexact.

Example 3.4. Which of the following quantities represent exact numbers? (a) The density of water at 70°C is 0.97778 g/mL. (b) There were 14 people riding the bus on the way in to work today. (c) The atmospheric pressure inside the eye of a hurricane is about 660 torr, or 660 mmHg. (d) There are 1,000,000 micrometers (microns) in a meter. (e) A human hair is about 150 μm wide.

Solutions. (a) Since density is a measured quantity, the value for the density of water is not exact. (b) Since people come in unit increments, you can say that you have exactly 14 people riding the bus, and that number is exact. (c) Atmospheric pressure is a measured property, so a pressure of 660 mmHg (millimeters of mercury) is not an exact number. (d) Because of the prefix system in SI units, the number of micrometers in a meter is a defined quantity, so the 1,000,000 can be considered an exact number. (e) The width of a human hair is a measured quantity and so is not exact.

The rule for significant figures in multiplication and division is not based on position, but on the number of significant figures of the quantities you are multiplying and dividing. In general, **your answer should be limited to the lowest number of significant digits of the values that are used to determine your answer.** Because exact numbers are considered to have an infinite number of sig figs, they do not affect the number of significant figures of the final answer.

For example, if you want to solve for V in the following expression:

$$(1.05)(V) = (0.08205)(298.15)$$

then you would divide both sides of the equation by 1.05. This isolates the variable V all by itself in the numerator of one side of the equation:

$$V = \frac{(0.08205)(298.15)}{1.05} = 23.29829285....$$

where the fraction has been properly evaluated, and the number given is what should appear on your calculator if you did the multiplication and division properly. In order to determine the proper number of significant figures in the answer, we need to determine the number of sig figs in the numbers we used to determine our answer. The number 0.08205 has 4 significant figures. (Do you agree with that?) 298.15 has 5 sig figs, and 1.05 has 3 significant figures. Of these, the least number of significant figures is 3 (from the 1.05), and the rule is to limit the final answer to the least number of significant figures. Therefore we limit our final answer to 3 significant figures. Our final answer, after the proper rounding, is

$$V = 23.3$$

Example 3.5. Evaluate the following expressions, limiting the final answer to the proper number of significant figures. (a) $(68.3)(0.003097)(2)$; the 2 is exact. (b) $-\dfrac{(8.314)(298.15)}{96,500}$ (c) $(0.0650)(6210)$

(d) $\dfrac{223.0}{71.0}$ (This is sometimes used as an approximation for π.)

Solutions. (a) The calculator answer is 0.4230502, but we have to limit it to three significant figures, so our final answer is 0.423. (b) The calculator answer is $-2.56872446 \times 10^{-2}$, but we must limit the answer to three significant figures. Therefore, the final answer is -2.57×10^{-2}. Notice that we had to round up, and that the 10^{-2} does not affect, or is not affected by, the significant figure determination. (c) The calculator answer is 403.65, but we limit the answer to three significant figures and get 404. (d) The calculator gives 3.14084507042.... as the answer, but we need to limit the answer to three significant figures, so we give 3.14 as the final answer.

What about problems that have addition and subtraction **and** multiplication and division? Probably the best tactic is to determine the proper number of significant figures for each part of the problem that you evaluate, using the appropriate rules. For example, suppose you need to evaluate the expression

$$4.32 - 56.92 \times (22.87 - 22.73)$$

Keeping in mind the general rule "PRMDAS," we will evaluate the expression inside the parentheses first, then perform the multiplication, then subtract that number from 4.32. So, for the first step, we evaluate

$$(22.87 - 22.73) = 0.14$$

Notice that this subtraction took us from four significant figures to only two! According to the order of operations, now the multiplication is evaluated. This means we must multiply 56.92 by the evaluated parenthetical expression.

$$56.92 \times 0.14 = 7.9688 = 8.0$$

We have limited 7.9688 to two significant figures, as required. Finally, we subtract 8.0 from 4.32.

$$4.32 - 8.0 = -3.68 = -3.7$$

We limit the significant figures to the tenths' place, since this is the rightmost common significant digit for the two numbers. Our final answer is -3.7, which has two significant figures.

Usually, you can keep all non-significant figures until your final answer. This means that you should not apply rules for sig figs when you perform the multiplication and division until after the addition and subtraction are performed. Your final answer should be very close to the answer, if not even the same answer, that you get if you limit your intermediate answers to the proper significant figures. If, for example, we performed all of the above operations and then applied the strictest sig fig rule (in this case, limitation of two sig figs from the multiplication operation), we would get

$$4.32 - 56.92 \times (22.87 - 22.73) = -3.6488 = -3.6$$

as our final answer. Notice that this differs from our earlier by one in the last significant digit. This is an example of truncation error, which is the slight variation of your final answer caused by imposing limits on the number of significant figures at different points in the calculation. Is this a big problem? No, it isn't. Remember that sig figs are those digits that are known, **plus the first uncertain digit.** Truncation errors are a natural part of significant figures, and their existence has influenced the rules by which significant figures are determined.

Significant figures are really an issue of common sense. Don't report more significant digits, numbers that are supposed to mean something, beyond the ability of the measurement device or beyond the significance of the numbers you use to do your calculations. We made a point in earlier material that most calculators do not work out the units. Well, most calculators don't care about significant figures either. It's up to the people working with the numbers to decide how many of the digits are significant.

Student Exercises

3.1. How long is each rod? Assume the units are centimeters.

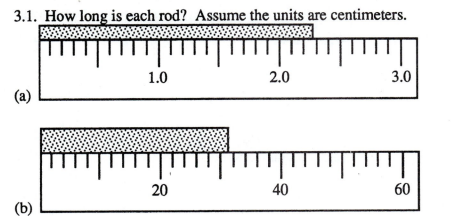

3.2.Read the following gauges to the proper number of significant figures.

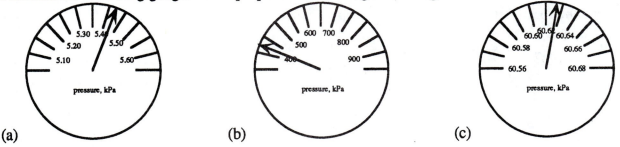

3.3. How many significant figures are in the following numbers? (a) 5.009×10^5 (b) 0.006510 (c) 1.000001 (d) -9.01.

3.4. How many significant figures are in the following numbers? (a) 0.00005 (b) 978,000 (c) 6.022×10^{23} (d) $6.6260755 \times 10^{-34}$.

3.5. Evaluate the following expressions to the proper number of significant figures. (a) 55,993 + 354.9 (b) 24,080 - 35 (c) 0.00103 - 0.00088 (d) 7.22 x 10^3 + 6.89 x 10^2

3.6. Evaluate the following expressions to the proper number of significant figures. (a) 102,993 + 6700 + 12,065 (b) 0.9634 - 0.0622 - 0.029 (c) 7845 + 3460 - 22.4

3.7. Would you consider π an exact number or not? Explain your answer.

3.8. Evaluate the following expressions to the proper number of significant figures. (a) 22.4 x 8.314 x 298.15 (b) $\dfrac{4.184 \cdot 2.08}{1.987}$ (c) $2 \cdot \dfrac{(1.61)^2}{4 \cdot \pi \cdot (6.626)^2}$ (assume that the 2, 4, and the π are exact).

3.9. Given the following expression

$$\frac{(745)(V)}{310.5} = \frac{(802)(4.50)}{298.15}$$

solve for V to the correct number of significant figures.

3.10. Given the following expression

$$\frac{(603.65)(25.88)}{598.2} = \frac{(499.20)(12.0)}{T}$$

solve for T to the correct number of significant figures.

3.11. Given the following expression

$$\Delta G = -(3)(96,477)(-1.019)$$

solve for ΔG to the correct number of significant figures. The 3 is exact.

3.12. Evaluate the following expressions. Impose the rules for significant figures different ways for the same expression. Do you find any truncation errors?

(a) 99.0 x (44.1 - (-22.007))

(b) 0.633 ÷ 21.9 + (3.40 x 0.09022)

(c) 1.0 ÷ 1.0 + (1.0 ÷ 7.5)

(d) $\dfrac{(681)(432 + 98.22)}{(-1005)(2.00 \div 0.00446)}$

Answers to Student Exercises

3.1. (a) About 2.28 cm long. (b) About 31.2 cm long.

3.2. (a) About 5.42 kPa. (b) About 420 kPa. (c) About 60.628 kPa

3.3. (a). 4 sig figs (b) 4 sig figs (c) 7 sig figs (d) 3 sig figs.

3.4. (a) 1 sig fig (b) 3 sig figs (c) 4 sig figs (d) 8 sig figs.

3.5. (a) 56,348 (b) 24,050 (rounding up) (c) 0.00015 (d) 7.91×10^3, or 7910.

3.6. (a) 121,800 (b) 0.872 (c) 11,280.

3.7. If all of the digits of π were used in a calculation, it should be considered as an exact number. If you only use a few digits, it is not exact. Simply using π as a variable implies it is exact, but using 3.14 as an approximation makes it not exact.

3.8. (a) 55,500 (b) 4.38 (c) 9.40×10^{-3}.

3.9. 5.04

3.10. 229

3.11. 294, 900, or 2.949×10^5

3.12. (a) 6540 (b) 0.336 (c) 1.1 (d) -0.802 if sig fig limits applied at each step in parentheses, -0.801 if sig fig rules (limit to three figures) applied only at the end of calculation.

Chapter 4. Converting Units

Introduction

In many chemistry calculations, you will be required to change units from one kind to another. This action is called several different things, and your book may prefer one name over another: dimensional analysis, the factor-label method, or, the preference here, <u>converting units</u>. Unit conversion is one of the most useful and widely applicable mathematical tools. In fact, we will extend this idea in the next chapter and see just how powerful a method it is.

Making Fractions from Equalities

We will start with the idea that 1 meter is equal to 100 centimeters. That is the definition of centimeter. We can write this statement as an equation:

$$1 \text{ meter} = 100 \text{ centimeter}$$

$$1 \text{ m} = 100 \text{ cm}$$

where in the second equation, we are using the abbreviations of the units. We should have an intuitive understanding that 100 cm **is** 1 m, so the above mathematical equation is acceptable to us.

Now, remember the rule in math that says that if you do something (i.e. add, subtract, multiply, divide) to both sides of an equation, what you generate is still an equation. That is, the left side still equals the right side of the equation. You have to perform the same operation **to both sides**. In this case, let us divide both sides of the equation by 100 centimeters. (And why not? There's no mathematical rule that forbids us to divide an expression with something that has a unit.) We get

$$\frac{1 \text{ m}}{100 \text{ cm}} = \frac{100 \text{ cm}}{100 \text{ cm}}$$

On the right-hand side of the equation, we have the same thing in the numerator and the denominator of the fraction, so it cancels. Not only the number cancels, but the unit cancels as well. Remember that we treat units algebraically, just like we treat numbers:

$$\frac{1 \text{ m}}{100 \text{ cm}} = \frac{100 \text{ cm}}{100 \text{ cm}} = 1$$

Because everything in the numerator and denominator cancels, we are left with 1. Therefore,

$$\frac{1 \text{ m}}{100 \text{ cm}} = 1$$

The value of the fraction $\frac{1 \text{ m}}{100 \text{ cm}}$ is exactly 1. This might not be as intuitively obvious as our original equation (1 m = 100 cm), but it makes sense. One meter is 100 centimeters, and a fraction that has the same quantity in the numerator and denominator is equal to one. It's just that in this fraction, the numerator and denominator are expressed in different units.

Keep in mind that we can also divide the original equation by 1 meter instead of 100 centimeters:

$$1 \text{ m} = 100 \text{ cm}$$

$$\frac{1 \text{ m}}{1 \text{ m}} = \frac{100 \text{ cm}}{1 \text{ m}}$$

$$\frac{1 \text{ m}}{1 \text{ m}} = \frac{100 \text{ cm}}{1 \text{ m}} = 1$$

$$\frac{100 \text{ cm}}{1 \text{ m}} = 1$$

and we get another fraction that equals one. You should be able to see that $\frac{100 \text{ cm}}{1 \text{ m}}$ is the reciprocal of $\frac{1 \text{ m}}{100 \text{ cm}}$, and the reciprocal of 1 is still 1. We therefore have two expressions that equal 1.

What we need to keep in mind is that **when you multiply a quantity by 1, that quantity does not change.** For example, 5 x 1 is still 5, and 25 meters times 1 is still 25 meters. The absolute value of the quantity does not change.

However, we can use the fractions that are equal to one to convert the units of an amount of something into different units. The absolute quantity remains the same, but the units we use to express that quantity are changed. We do this by expressing an equivalence as a fraction, putting the unit we want

from, and its associated number, in the **denominator** of the fraction. The unit (and its associated number) we want to convert **to** is in the numerator. We multiply this fraction, which equals 1, by our initial quantity. Then, we multiply and divide our numbers and get a new value with a new unit – but it's still the same quantity.

For example, we know that 50 centimeters is half of a meter, or 0.5 meters. Mathematically, however, we can set up a conversion like this:

$$50 \text{ cm} \cdot \frac{1 \text{ m}}{100 \text{ cm}}$$

where the fraction $\frac{1 \text{ m}}{100 \text{ cm}}$ equals 1, and so by multiplying 50 cm by this fraction, we are making a new expression that is equivalent to the original quantity "50 cm." Remember that we can treat the 50 cm as a numerator with 1 in the denominator. What this means is that we have the product of two fractions **with the unit of cm in both numerator and denominator.** Therefore, the unit "cm" cancels:

$$50 \text{ c\!\!\!/m} \cdot \frac{1 \text{ m}}{100 \text{ c\!\!\!/m}}$$

When we perform the remaining multiplications and divisions, we get

$$50 \cdot \frac{1 \text{ m}}{100} = 0.50 \text{ m}$$

where our final answer now has units of meters. The quantity has not changed its absolute **value**, it has only changed its **units**.

There are two simple hints for converting units:

- **Start with something that you are given.** In a majority of mathematical problems, this rule works. In problems, you will be given some data, some quantities. In most cases, the problem can be solved by taking one of the quantities and converting it to another unit.

- **The unit you want to get rid of usually goes in the denominator of the conversion factor you multiply the quantity by.** In doing this, we use the normal rules of algebra and cancel units from numerator and denominator.

As a counterexample, suppose we were to multiply our 50 cm by the other fraction that equals 1, the $\frac{100 \text{ cm}}{1 \text{ m}}$:

$$50 \text{ cm} \cdot \frac{100 \text{ cm}}{1 \text{ m}}$$

This is OK, isn't it? After all, we are only multiplying our 50 cm by 1 again. But in this case, the **units** don't work out: the cm unit does not cancel algebraically, and we get a nonsensical expression if we try to multiply and divide everything through. **Use the units to help you construct proper conversion factors and it will work out correctly every time.** If you don't pay attention to the units in these conversion factors, you will quickly go wrong. Keeping track of the units can actually help you work out a problem.

Example 4.1. Set up and evaluate the following conversions algebraically. (a) 0.75 m to cm. (b) 3.66 km to m. (c) 1855 mL to L. (d) 3.88 x 10^4 g to kg.

Solutions. The following are the algebraic constructions for performing the conversions. Notice that they all start with the quantity that is given. You might want to plug the numbers into your calculator to verify the final numerical result, and write on the page to show how the units cancel.

(a) $0.75 \text{ m} \cdot \dfrac{100 \text{ cm}}{1 \text{ m}} = 75 \text{ cm}$ (b) $3.66 \text{ km} \cdot \dfrac{1000 \text{ m}}{1 \text{ km}} = 3660 \text{ m}$

(c) $1855 \text{ mL} \cdot \dfrac{1 \text{ L}}{1000 \text{ mL}} = 1.855 \text{ L}$ (d) $3.88 \text{ x } 10^4 \text{ g} \cdot \dfrac{1 \text{ kg}}{1000 \text{ g}} = 3.88 \text{ x } 10^1 \text{ kg} = 38.8 \text{ kg}$

More Than One Conversion

Of course, more than one conversion can be applied to a quantity. In that case, each conversion can be performed sequentially. For instance, in converting 0.0678 km into millimeters, we will first convert the km units into units of meters:

$$0.0678 \text{ km} \cdot \frac{1000 \text{ m}}{1 \text{ km}} = 67.8 \text{ m}$$

where the kilometer units in the numerator and denominator are canceled algebraically. We now convert the meters units into millimeters, canceling the meter units, for our final answer. Any number of proper conversions can be performed sequentially like this.

Of course, we could have performed this conversion in a single step if we recognized that there are 1,000,000 mm in 1 km. But instead of memorizing all possible conversion factors between the prefixed units (like how many microliters there are in a gigaliter), **it is usually easier to first convert from the the original prefixed unit to the basic (unprefixed) unit, and from the basic unit to the final prefixed unit.** In the above case, we went from kilometers (a prefixed unit, with the prefix being "kilo-") to meters (the basic unit) and then from meters to millimeters (the final prefixed unit, with "milli-" being the prefix this time).

Example 4.2. Perform the following conversions in two steps by first converting to the basic, unprefixed unit in each case. (a) 5.7×10^{-5} kW into μW (W = watts) (b) 9.992×10^{12} mHz into MHz (where the unit hertz, Hz, is a unit of frequency and equals $\dfrac{1}{\text{sec}}$).

Solutions. (a) To go from kilowatts to microwatts, we first convert our units into the basic unit of watts,

$$5.7 \times 10^{-5} \, \cancel{\text{kW}} \cdot \frac{1000 \text{ W}}{1 \, \cancel{\text{kW}}} = 5.7 \times 10^{-2} \text{ W}$$

and then convert the watts into microwatts,

$$5.7 \times 10^{-2} \, \cancel{\text{W}} \cdot \frac{10^6 \, \mu\text{W}}{1 \, \cancel{\text{W}}} = 5.7 \times 10^4 \, \mu\text{W}$$

Our final answer is $5.7 \times 10^4 \, \mu$W. For (b), it is a similar process except we are converting millihertz into megahertz. The first step is to convert the millihertz to the basic unit of hertz,

$$9.992 \times 10^{12} \, \cancel{\text{mHz}} \cdot \frac{1 \text{ Hz}}{1000 \, \cancel{\text{mHz}}} = 9.992 \times 10^9 \text{ Hz}$$

and now from Hz to MHz,

$$9.992 \times 10^9 \, \cancel{Hz} \cdot \frac{1 \text{ MHz}}{10^6 \, \cancel{Hz}} = 9.992 \times 10^3 \text{ MHz} = 9992 \text{ MHz}$$

In both cases, the unit cancellations are shown.

After performing a lot of unit conversions, you might begin to feel that they are all very similar, and they are. The same mathematical techniques are used for all unit conversions, so when you master those techniques, you can do any conversion problem.

After a bit of sophistication in these conversion problems is developed, we can take the process to another level. Instead of performing multi-step conversions in separate parts, **we can group the steps into a single, longer step.** Mathematically, our final answer will be exactly the same as if we performed the conversion in steps – as long as we include all of the proper steps.

We will convert 0.0678 km into mm again, but this time in one long step:

$$0.0678 \text{ km} \cdot \frac{1000 \text{ m}}{1 \text{ km}} \cdot \frac{1000 \text{ mm}}{1 \text{ m}} = 67,800 \text{ mm}$$

We get the same answer as before. Notice how the units work out: the first fraction cancels the "km" unit,

$$0.0678 \, \cancel{\text{km}} \cdot \frac{1000 \text{ m}}{1 \, \cancel{\text{km}}} \cdot \frac{1000 \text{ mm}}{1 \text{ m}} = 67,800 \text{ mm}$$

and the second fraction cancels the "m" unit,

$$0.0678 \, \cancel{\text{km}} \cdot \frac{1000 \, \cancel{\text{m}}}{1 \, \cancel{\text{km}}} \cdot \frac{1000 \text{ mm}}{1 \, \cancel{\text{m}}} = 67,800 \text{ mm}$$

The only unit left is millimeter, which is the unit we are converting into. In plugging the numbers into a calculator, we could either hit the = or **ENTER** or **EXECUTE** key after each number, or just use the **X** key before entering any number in a numerator and the ÷ key before any number in a denominator. Our sequence of keys would then be

0.0678 X 1000 X 1000 =

(where we are not including the 1's in the denominators).

Finally, we recognize the mathematical equivalence of this method and doing the conversion in steps by pointing out that the first two terms, the 0.0678 km and the first fraction, are collectively **equal to 67.8 m**:

$$0.0678 \text{ km} \cdot \frac{1000 \text{ m}}{1 \text{ km}} \cdot \frac{1000 \text{ mm}}{1 \text{ m}} = 67,800 \text{ mm}$$

$$\underbrace{\qquad\qquad\qquad} = 67.8 \text{ m}$$

In a sense, all we are doing is a substitution of the first step into the second step of this conversion. In doing so, we accomplish the complete conversion all at once instead of a step at a time.

Example 4.3. Re-do the two problems in Example 4.2, but in a single step.

Solutions. You should be able to convince yourself that the two examples will look like this:

(a)
$$5.7 \times 10^{-5} \text{ kW} \cdot \frac{1000 \text{ W}}{1 \text{ kW}} \cdot \frac{10^6 \text{ } \mu\text{W}}{1 \text{ W}} = 5.7 \times 10^4 \text{ } \mu\text{W}$$

(b)
$$9.992 \times 10^{12} \text{ mHz} \cdot \frac{1 \text{ Hz}}{1000 \text{ mHz}} \cdot \frac{1 \text{ MHz}}{10^6 \text{ Hz}} = 9.992 \times 10^3 \text{ MHz} = 9992 \text{ MHz}$$

In both cases, we get the same answer. For (b), the sequence of keys in your calculator might be

9.992 EE 12 ÷ 1000 = ÷ 1 EE 6 =

The first = evaluates the division from the first conversion factor, and the very next ÷ sign indicates that the next number, which is really 1×10^6, is also in the denominator and should be dividing the expression. **You should practice putting these numbers in your calculator and verify that you get the same answer given here.** If you do not, you may not be using your calculator correctly.

Density: A New Type of Conversion Factor

The density of an object is defined as the mass of the object divided by its volume:

$$density = d = \frac{mass}{volume}$$

Density usually has units of $\frac{g}{mL}$, $\frac{g}{cm^3}$, or even $\frac{g}{L}$ if the material is a gas. These units are read as "grams per milliliter," "grams per cubic centimeter," and "grams per liter," respectively. For example, the density of iron metal is $7.86 \ \frac{g}{cm^3}$.

Remember that fractions can be thought of as having a "1" in the denominator. Therefore, the density of iron can be written as

$$\frac{7.86 \ g}{1 \ cm^3}$$

which is read as "7.86 grams per one cubic centimeter." That is, for every cubic centimeter of volume, a piece of iron has a mass of 7.86 grams. Another way of writing this is as an equation. We say that, for iron,

$$7.86 \ g = 1 \ cm^3$$

Density, therefore, represents a conversion factor between mass and volume. If we have a mass, we can use density to determine the volume of that mass. If we have a volume, we can use density to determine the mass of that volume.

There are three things to keep in mind. First, different materials have different densities. That's why we specified that the above density is "for iron." The above density is for iron only. It would not be correct to use this value for the density if your sample is, say, water or balsa wood or sodium chloride. To keep this in mind, it may be better to write the mass-volume relationship above as

$$7.86 \ grams \ of \ iron = 1 \ cm^3 \ of \ iron$$

to emphasize that this equivalence is applicable only to iron. (We will use memory devices like this a lot in the next chapter.) In working a problem, you must use the correct density for the material being considered. Second, your mass and volume units must be consistent when working out a mathematical problem! If your density is given in terms of $\frac{g}{mL}$ and your volume is in units of liters, you must convert a volume unit so that it will appropriately cancel out in order to determine the mass of your sample.

Third, recognize that you can use the above equation to make two different conversion factors. The first is the normal definition of density:

$$\frac{7.86 \text{ g of iron}}{1 \text{ cm}^3 \text{ of iron}}$$

This is useful for volume-to-mass conversions. You can **also** put the <u>mass</u> in the denominator:

$$\frac{1 \text{ cm}^3 \text{ of iron}}{7.86 \text{ g of iron}}$$

This is mathematically the reciprocal of the density, and can be used for mass-to-volume conversions. The density thus acts as two different conversion factors. Which one do you use? It depends on what unit you are trying to cancel and what unit you ultimately want for your final answer.

Example 4.4. These examples all use density as a conversion factor and go from simple to more complex. (a) What is the mass of 11.83 cm^3 of iron metal? (b) What is volume of 100.0 grams of silver nitrate, AgNO$_3$, which has a density of 4.35 grams per cubic centimeter? (c) The airship *Hindenberg*, which exploded in 1937, was filled with 2.000×10^{11} mL of H$_2$ gas. At a density of 0.0824 grams per liter, how many kilograms of hydrogen is this?

Solutions. (a) In order to convert from volume to mass, we use the density conversion factor that has volume in the denominator. This corresponds to the original definition of density:

$$11.83 \cancel{\text{ cm}^3 \text{ of iron}} \cdot \frac{7.86 \text{ grams of iron}}{1 \cancel{\text{ cm}^3 \text{ of iron}}} = 93.0 \text{ grams of iron}$$

where we limit our final answer to 3 significant figures. Notice how the "cm^3 of iron" units cancel. (b) To convert from mass to volume, we use the inverse of density because we want the units of mass to cancel:

$$100.0 \text{ g of AgNO}_3 \cdot \frac{1 \text{ cm}^3 \text{ of AgNO}_3}{4.35 \text{ g of AgNO}_3} = 23.0 \text{ cm}^3 \text{ of AgNO}_3$$

Again, the final answer is limited to three significant figures. (c) In order to determine the mass of hydrogen, we must make the volume units consistent with each other. We can either convert the volume of H_2 to liters, or the density of H_2 to grams per mL. The answer will be the same either way the problem is worked. Let us convert the initial volume to units of liters, then use the density as a conversion factor to determine the number of grams of H_2 we have. Finally, we will include one more conversion factor to convert the mass in grams to a mass in units of kilograms:

$$2.000 \times 10^{11} \text{ mL} \cdot \underbrace{\frac{1 \text{ L}}{1000 \text{ mL}}}_{\substack{\text{converts} \\ \text{to liters}}} \cdot \underbrace{\frac{0.0824 \text{ g}}{1 \text{ L}}}_{\substack{\text{converts} \\ \text{to mass}}} \cdot \underbrace{\frac{1 \text{ kg}}{1000 \text{ g}}}_{\substack{\text{converts} \\ \text{to kg}}} = 1.65 \times 10^4 \text{ kg}$$

Each step in this problem, the most complicated one we have worked out so far, cancels a unit and introduces a new one. The final conversion gives us the unit we need, kg. You should practice plugging the numbers into your calculator for each example – especially the last one! – and verify that you can reproduce the answer that is given.

Other Conversion Factors

There are several other quantities defined in science and chemistry that are similar to density in that they contain two (or more) different types of units and therefore represent conversion factors between those types of units. Density is the conversion factor between mass and volume. Other quantities that can be used mathematically **just like density** are listed on the next page.

Quantity	Abbreviation	Definition	Types of Units
moles	mol	1 mole = 6.02×10^{23} atoms or molecules	amount and atoms/molecules
moles	mol	1 mole = 1 molecular weight of material, in grams	amount and mass
molarity	M	$\dfrac{\text{\# moles of solute}}{\text{\# L of solution}}$	moles and volume of solution
molality	m	$\dfrac{\text{\# moles of solute}}{\text{\# kg of solvent}}$	moles and mass of solvent
energy per mole	$\dfrac{kJ}{mol}$	$\dfrac{\text{amount of energy}}{\text{mole of material}}$	energy and amount
rate of reaction	$\dfrac{mol}{sec}, \dfrac{M}{sec}$	$\dfrac{\text{change in amount}}{\text{change in time}}$	amount and time

In all cases, the abbreviation represents a new unit that, with a number, can be used to express the related quantity. (For example, "0.25 M" is a concentration that is equal to 0.25 moles of solute per liter of solution.) For each quantity, either there is an equation in the definition that can be turned into a conversion factor (like the first two entries) or the definition is written as a fraction that, like density, can be used directly to go from one type of unit to another (i.e. the last four entries). In both cases, the quantities can be used algebraically just like we used density in the previous section. In the next chapter, we will begin to use these quantities quite a bit when we work out mathematical problems.

All of the new units above are related to <u>amount</u>. In chemistry, amount of material is a very important quantity and its unit, the <u>mole</u>, is perhaps the most useful unit of chemistry.

Converting Combined Units

Suppose we have a unit that is a combination of several fundamental units. How do we convert those? What you do is **break apart the combined unit into the product of its individual units and multiply by a conversion factor for each individual unit.**

For example, we have defined a newton as a $\dfrac{kg \cdot m}{sec^2}$. There is another unit of force called the <u>erg</u>, which has the fundamental unit definition of $\dfrac{g \cdot cm}{sec^2}$. How many ergs are there in one newton?

We apply a conversion factor for each unit that we need to change. First, let us convert the kg unit to g:

$$1 \frac{\cancel{kg} \cdot m}{sec^2} \cdot \frac{1000 \text{ g}}{1 \cancel{kg}} = 1000 \frac{g \cdot m}{sec^2}$$

Then, we can convert the m to cm (of course, the order of conversions does not matter; you could do m to cm first, then kg to g):

$$1000 \frac{g \cdot \cancel{m}}{sec^2} \cdot \frac{100 \text{ cm}}{1 \cancel{m}} = 100{,}000 \frac{g \cdot cm}{sec^2} = 10^5 \text{ ergs}$$

There are 100,000 ergs in one newton. Similar conversions can be performed just as easily.

Example 4.5. The liter was originally defined as a cubic decimeter, dm^3. A decimeter is one-tenth of a meter. How many cubic centimeters, cm^3, are there in a liter?

Solution. This example requires several conversions. We will start with the fact that

$$1 \text{ L} = 1 \text{ dm}^3$$

and convert to units of meters, and then to units of centimeters. We will rewrite the cubic decimeter as the explicit product of three decimeter units:

$$1 \text{ dm}^3 = 1 \text{ dm} \cdot dm \cdot dm$$

It will be easier to see the conversion this way. Using the rules for converting combined units, we will need to convert each dm to m, which means that we will have to use the same conversion factor **three times**:

$$1 \cancel{dm \cdot dm \cdot dm} \cdot \frac{1 \text{ m}}{10 \cancel{dm}} \cdot \frac{1 \text{ m}}{10 \cancel{dm}} \cdot \frac{1 \text{ m}}{10 \cancel{dm}} = 0.001 \text{ m}^3$$

Now we can convert the m units to cm units by again writing the m^3 as $m \cdot m \cdot m$ and applying the appropriate conversion factor three times:

$$0.001 \; \cancel{\text{L·L·L}} \cdot \frac{100 \text{ cm}}{1 \; \cancel{\text{L}}} \cdot \frac{100 \text{ cm}}{1 \; \cancel{\text{L}}} \cdot \frac{100 \text{ cm}}{1 \; \cancel{\text{L}}} = 1000 \text{ cm}^3$$

There are 1000 cm³ in one liter. Since there are also 1000 mL in one liter, we establish the fact that

$$1 \text{ mL} = 1 \text{ cm}^3$$

where both 1's can be considered exact numbers (so they don't affect the significant figure determination).

Student Exercises

Try to express your final answers in the proper number of significant figures.

4.1. Construct two conversion factors from the following relationships.

(a) 1 inch = 2.54 cm

(b) 1 dozen = 12 things

(c) 1 m = 10^6 μm

(d) 1 cm³ = 1 mL

4.2. Convert, in a single step:

(a) 5.209 kW into W

(b) 20.0 μL into L

(c) 38,000 g into Mg

(d) 0.000 000 035 torr into mtorr (torr is a unit of pressure)

4.3. Convert, in two steps:

(a) 35.4 MHz to mHz

(b) 7.24 x 10^7 nm to km

(c) 29.11 µL to mL   (d) 0.95 msec to nsec

4.4. The human eye is most sensitive to green light, which has a wavelength of about 5500 Å, where 1 Å ('Ångstrom') equals 1×10^{-10} m. What is the wavelength of green light in units of m?

4.5. Evaluate the following expressions using your calculator, and determine the correct final unit:

(a) $78.3 \text{ kg} \cdot \dfrac{1000 \text{ g}}{1 \text{ kg}} \cdot \dfrac{1000 \text{ mg}}{1 \text{ g}} =$

(b) $0.073 \text{ L} \cdot \dfrac{832 \text{ g}}{1 \text{ L}} \cdot \dfrac{1 \text{ mol}}{40.04 \text{ g}} =$

(c) $1.088 \text{ L} \cdot \dfrac{1 \text{ mol}}{22.4 \text{ L}} \cdot \dfrac{28.02 \text{ g}}{1 \text{ mol}} \cdot \dfrac{1000 \text{ mg}}{1 \text{ g}} =$

4.6. What is the density of a material, in units of $\dfrac{g}{cm^3}$, if 0.497 kg has a volume of 0.107 L? HINT: Use the answer in Example 4.5 to help with your unit conversion.

4.7. What mass of helium has a volume of 125.0 L if the density of helium gas is $0.1787 \dfrac{g}{L}$?

4.8. Sodium chloride, NaCl, has a density of 2.17 $\frac{g}{cm^3}$. If you needed 250.0 grams of NaCl, what volume would you need?

4.9. Which has more mass, 2750 mL of argon gas having a density of 1.784 $\frac{g}{L}$ or 4.50 cm^3 of phenol, which has a density of 1.058 $\frac{g}{cm^3}$? (Argon gas is sometimes used to fill light bulbs; phenol is the active ingredient in some sore-throat lozenges.)

4.10. How many mm^3 are there in one liter?

4.11. How many atoms are there in 1.00 grams of xenon gas? Xenon has at atomic weight of 131.3 $\frac{grams}{mole}$. HINT: You will have to use both definitions of the unit "mole" in a two-step conversion.

4.12. If a chemical reaction gives off 383.51 kJ of energy per mole of carbon dioxide produced as a product, how much energy is this in units of calories per gram of CO_2 produced? The molecular weight of CO_2 is 44.00 grams per mole, and there are 4.184 J in 1 calorie. (This is an exact conversion).

4.13. What volume of solution, in liters, is necessary to obtain 0.250 moles of solute from a solution whose concentration is given as 0.1059 M?

4.14. If you are given a sodium chloride solution that is 2.054 M in concentration, what volume of solution is needed to get 25.0 grams of dissolved NaCl? The formula weight of NaCl is 58.44 grams per mole.

Answers to Student Exercises

4.1. (a) $\dfrac{1\text{ inch}}{2.54\text{ cm}}$ and $\dfrac{2.54\text{ cm}}{1\text{ inch}}$ (b) $\dfrac{1\text{ dozen}}{12\text{ things}}$ and $\dfrac{12\text{ things}}{1\text{ dozen}}$ (c) $\dfrac{1\text{ m}}{10^6\ \mu m}$ and $\dfrac{10^6\ \mu m}{1\text{ m}}$

(d) $\dfrac{1\text{ cm}^3}{1\text{ mL}}$ and $\dfrac{1\text{ mL}}{1\text{ cm}^3}$

4.2. (a) 5209 W (b) 0.000 020 0 L, or 2.00×10^{-5} L (c) 0.038 Mg, or 3.8×10^{-2} Mg

(d) 0.000 035 mtorr, or 3.5×10^{-5} mtorr.

4.3. 35,400,000,000 mHz, or 3.54×10^{10} mHz (b) 0.0724 km, or 7.24×10^{-2} km (c) 0.02911 mL, or 2.911×10^{-2} mL (d) 950 nsec, or 9.5×10^2 nsec.

4.4. 0.000 000 55 m, or 5.5×10^{-7} m.

4.5. (a) 78,300,000 mg, or 7.83×10^7 mg (b) 1.5 mol (c) 1360 mg, or 1.36×10^3 mg.

4.6. 4.64 $\dfrac{g}{cm^3}$.

4.7. The helium would have a mass of 22.34 grams.

4.8. 115 cm^3 of NaCl would be needed.

4.9. The argon would have a mass of 4.91 grams, and the phenol would have a mass of 4.76 grams. The argon would therefore have slightly more mass.

4.10. There are 1,000,000 (or 1×10^6) mm^3 in one liter.

4.11. There are 4.58×10^{21} atoms of xenon in 1.000 gram of xenon.

4.12. The reaction gives off 2083 calories for every gram of CO_2 produced.

4.13. You will need 2.36 liters of solution to get 0.250 moles of solute.

4.14. You will need 0.208 liters of solution to get 25.0 grams of dissolved sodium chloride from the solution.

Chapter 5. Using Chemical Reactions to Make Conversion Factors

Introduction

One of the central themes in chemistry is the <u>balanced chemical reaction</u>. It provides, in a nutshell, a statement of what initial chemical substances are changing into what final chemical substances. Respectively, these substances are called <u>reactants</u> and <u>products</u>. Because of the Law of Conservation of Mass, the amount of mass of reactants must equal the amount of mass of products, and that's why we have to write **balanced** chemical reactions.

A balanced chemical reaction therefore contains **quantitative** information, both in terms of grams and moles of reactants and products. This quantitative information can be used to perform a variety of calculations. This chapter reviews some of the techniques for using balanced chemical reactions mathematically, by constructing conversion factors to use in calculations.

Balanced Chemical Reactions and Moles

Consider how chemistry summarizes the reaction of hydrogen gas, H_2, with oxygen gas, O_2, to make the product water, H_2O:

$$H_2 \,(g) \;+\; O_2 \,(g) \;\rightarrow\; H_2O \,(\ell)$$

We will drop the phase labels in the future, for reasons of clarity. The first step in writing any balanced chemical reaction is to start with proper formulas for the reactants and products. Both hydrogen and oxygen are diatomic elements, and the correct formula for water is well known.

If we consider this reaction at the atomic scale, then what we are saying is, "One molecule of hydrogen and one molecule of oxygen are reacting to make one molecule of water." This is the correct way of wording it, but there's a problem: it violates the Law of Conservation of Mass. If you count the number of hydrogen **atoms** as reactants, there are a total of two; and if you count the number of hydrogen atoms as products, there are also two. But there are two oxygen atoms as reactants and **only one** oxygen atom in the products! Where did the other oxygen atom go? One of the fundamental ideas in chemistry is that you must have the same number of atoms of each element as products and as reactants.

We cannot change the formulas of the individual reactants and products. However, we can change the number of molecules of reactants and products. This takes some trial-and-error and

73

experience, but in time it becomes a relatively easy task. Properly-balanced reactions typically use the lowest whole numbers necessary to make the number of atoms of each element the same on both sides of the reaction. For the reaction of hydrogen and oxygen to make water, the properly balanced reaction is

$$2\,H_2 + O_2 \rightarrow 2\,H_2O$$

You should check this to make sure there are the same number of atoms of each element on both sides of the reaction.

This balanced chemical reaction says, "Two molecules of hydrogen and one molecule of oxygen react to make two molecules of water." This is a proper interpretation of the balanced chemical reaction. But it is very difficult to follow reactions at the atomic or molecular scale. Atoms and molecules are just too small for us to follow them reacting one at a time. Chemistry defines a unit called the mole that solves this problem. A mole represents a certain number of things, just as a dozen represents a certain number of things. But whereas a dozen is 12, a mole represents a much larger number:

$$1 \text{ mole} = 6.02 \times 10^{23}$$

This number is called Avogadro's number, after an Italian chemist who proposed its existence. Why does a mole represent this number of things? Well, atoms have mass. In very small units called atomic mass units (or amu), one hydrogen atom has a mass of about 1.008 amu. One oxygen atom has a mass of about 16.00 amu, and an average mercury atom has a mass of about 200.6 amu. One **mole** of hydrogen atoms, 6.02×10^{23} H atoms, has the same amount of mass but in units of grams: one mole of hydrogen atoms has a mass of 1.008 **grams**. One mole of mercury atoms, 6.02×10^{23} Hg atoms, has a mass of 200.6 **grams**. The mole thus acts as the conversion from microscopic units (amu) to macroscopic units (grams).

The mole/mass relationship is also applicable to molecules. We simply add up the masses of the individual atoms in the formula, and one mole of that molecule has that sum of masses in grams. For instance, since there are two hydrogen atoms in a hydrogen molecule, a single H_2 molecule has a mass of $1.008 + 1.008 = 2.016$ amu, so a **mole** of hydrogen molecules has a mass of 2.016 grams. Note the difference: a mole of hydrogen **atoms** has a mass of 1.008 grams, but a mole of hydrogen **molecules** has a mass of 2.016 grams. It is important to start keeping track of the exact chemical identity of the material you are working with so you don't get confused. The following example shows how moles of atoms can be different from moles of molecules.

Example 5.1. How many moles of hydrogen atoms are there in (a) one mole of water, H_2O (b) 2.5 moles of benzene, C_6H_6 (c) one-half mole of hydrogen peroxide, H_2O_2.

Solutions. (a) Since the chemical formula for water shows that there are two atoms of hydrogen in every molecule of water, then in one mole of water there are a total of two moles of individual hydrogen atoms. (b) The formula for benzene indicates that there are six hydrogen atoms in every molecule of C_6H_6, so there are 6 x 2.5 = 15 moles of individual hydrogen atoms. (c) Like water, hydrogen peroxide has two hydrogen atoms in each molecule, so in one-half mole of H_2O_2 there are $2 \times \frac{1}{2} = 1$ mole of hydrogen atoms.

The mass of one mole of atoms of any element is called the <u>atomic weight</u> of that element. Atomic weights are typically listed in periodic tables. The mass of one mole of molecules, or one formula unit of an ionic compound, is called the <u>molecular weight</u> of that compound.* (<u>Formula weights</u> are sometimes used for ionic compounds.) We therefore have the relationship that

1 mole of material = 1 atomic or molecular weight of that material

We can use this equation to make a conversion factor between amount in moles and mass in grams. The next example shows how we do this.

Example 5.2. What is the mass of (a) 0.0550 moles of mercury, Hg (b) 2.75 moles of NaCl.

Solutions. (a) First, let us set up our conversion factors. Using the mole unit and its relationship to the atomic weight of an element:

$$1 \text{ mol Hg} = 200.6 \text{ g Hg}$$

Notice that we are being explicit about "moles of what" and "grams of what." It is important to keep track of this. To construct a conversion factor from the above equation, remember that we are given an amount in moles at the beginning, and that we want to cancel this out and convert to units of grams. We therefore make our conversion factor with the mole unit in the denominator:

---

* Although these terms use the word 'weight', they are actually masses. <u>Weight</u> is a force due to gravity; <u>mass</u> is an inherent property of matter. However, the terms are so embedded in science that it would be fruitless to try to change them.

$$\frac{200.6 \text{ g Hg}}{1 \text{ mol Hg}} = 1$$

This fraction **does** equal one, because one mole of mercury **is** 200.6 grams of mercury! That is the atomic weight of mercury. We use this conversion factor algebraically just like any other conversion factor. Starting out with what we are given, we now convert from moles to grams:

$$0.0550 \text{ mol Hg} \cdot \frac{200.6 \text{ g Hg}}{1 \text{ mol Hg}} = 11.0 \text{ g Hg}$$

Notice that the "mol Hg" cancels algebraically, like any other unit. Our final answer is limited to three significant figures. (b) In this case, we are dealing with a compound, so we need to determine the molecular weight of NaCl. Using the periodic table, we find that the atomic weight of sodium is 22.9898 and the atomic weight of chlorine is 35.453. Adding them together, we get 58.443 as the molecular weight for NaCl, which means that

$$1 \text{ mol NaCl} = 58.443 \text{ g NaCl}$$

Again, we want a conversion factor that allows us to cancel the "mol NaCl" unit in what we are given, so we use the above equation to construct the conversion factor

$$\frac{58.443 \text{ g NaCl}}{1 \text{ mol NaCl}}$$

Using this as a conversion factor, we get

$$2.75 \text{ mol NaCl} \cdot \frac{58.443 \text{ g NaCl}}{1 \text{ mol NaCl}} = 161 \text{ g NaCl}$$

This example shows two things. First, we can construct conversion factors to go from units of moles to units of grams, using atomic or molecular weights. Part (a) used an atomic weight, and part (b) used a molecular weight, but the conversion itself was similar in both cases. Second, we must be extremely careful to keep track of what material we are working with. For instance, in part (a) above we found that one mole was equal to 200.6 grams, but in part (b) we found that one mole was equal to 58.443 grams. **That's because we were working with different materials!** Every chemical has its own characteristic atomic or molecular weight, and it is absolutely essential that you keep track of what material you are referring to when you perform these types of calculations. That's why we included the "Hg" and "NaCl" when we used "mole" or "gram" units: it helps us keep track of what material we are talking about. In the next few sections, the need to keep track of what material is being referred to will become even more obvious.

What is the relationship between moles and balanced chemical reactions? **A balanced chemical reaction can be considered as written in terms of moles of reactants and products, not molecules.** Therefore,

$$2\,H_2 + O_2 \rightarrow 2\,H_2O$$

can be spoken as "2 moles of hydrogen and 1 mole of oxygen react to make 2 moles of water." All balanced chemical reactions can be thought of as being balanced on a **molar** scale, not just an atomic or molecular scale.

Occasionally, one sees a chemical reaction that is balanced using fractional coefficients. This is not considered absolutely wrong, but some textbooks do not use them as a matter of course. For instance,

$$H_2 + \frac{1}{2}\,O_2 \rightarrow H_2O$$

is considered balanced. But how can we speak of half of a molecule? We don't: we speak of one-half of a **mole** of diatomic oxygen, which equals one mole of oxygen atoms. Since there is one mole of oxygen atoms in the products, the chemical reaction is balanced. **You should check with your textbook and your instructor to find out if using fractional coefficients is acceptable in your course.**

Mole-Mole Problems

A balanced chemical reaction is a source of various conversion factors that we can use to make a variety of calculations. Again, consider the following reaction:

$$2\,H_2 + O_2 \rightarrow 2\,H_2O$$

If we ask ourselves, "How many moles of oxygen will react with 2 moles of hydrogen," the answer is obvious. One mole of oxygen reacts with 2 moles of hydrogen. Suppose we are going to react 20 moles of hydrogen? We will react 10 moles of oxygen. We are using the 2:1 ratio given by the coefficients in the balanced chemical reaction.

One way of looking at this is that **2 moles of hydrogen are chemically equivalent to one mole of oxygen.** That is, we can write

$$2 \text{ mol } H_2 = 1 \text{ mol } O_2$$

as far as the above reaction is concerned. (This seems an unusual "equation" to write, but the **balanced chemical reaction** says that hydrogen and oxygen react in this proportion. We think of the "equivalence" as a chemical one that we can take advantage of mathematically.) This allows us to construct the following conversion factors:

$$\frac{2 \text{ mol } H_2}{1 \text{ mol } O_2} = \frac{1 \text{ mol } O_2}{2 \text{ mol } H_2} = 1$$

We can use these conversion factors to calculate the moles of one substance that will react with a given number of moles of another substance.

Example 5.3. How many moles of O_2 will react with 7.33 moles of H_2 to make water?

Solution. We will start our calculation with what we are given: 7.33 moles of H_2. We can then use the proper conversion factor, the one with "mol H_2" in the denominator, to convert to moles of O_2:

$$7.34 \text{ mol } H_2 \cdot \frac{1 \text{ mol } O_2}{2 \text{ mol } H_2} = 3.67 \text{ mol } O_2$$

Notice how the "mol H_2" unit cancels in the example above. Suppose, however, the conversion factor was written simply

$$\frac{1 \text{ mol}}{2 \text{ mol}}$$

There might be a temptation to cancel the "mol" units within this conversion factor. Conceptually, this would be incorrect: the "mol" unit in the numerator refers to moles of **oxygen**, whereas the "mol" unit in the denominator refers to moles of **hydrogen**. It is important that you are aware of this difference so you do not get your units confused. That's why we write "mol H_2" and "mol O_2," and you are encouraged to identify your units properly to minimize making errors.

In the balanced chemical reaction, **all** of the products and reactants are chemically equivalent, so there are many possible conversion factors you can construct. For the reaction of hydrogen and oxygen making water, we have

$$2 \text{ mol } H_2 = 1 \text{ mol } O_2 = 2 \text{ mol } H_2O$$

so we can devise conversions not only between reactants but also between reactants and products.

Example 5.4. How many moles of water are produced when 4.89 moles of hydrogen gas reacts with oxygen?

Solution. Since we are dealing with hydrogen and water, we need a conversion factor that involves these two chemicals. From the balanced chemical reaction, we see that two moles of hydrogen react to make two moles of water. Since we are given an amount of hydrogen and want an amount of water, we put moles of hydrogen in the denominator of the conversion factor and moles of water in the numerator:

$$\frac{2 \text{ mol } H_2O}{2 \text{ mol } H_2}$$

Solving the problem, we start with what we are given:

$$4.89 \text{ mol } H_2 \cdot \frac{2 \text{ mol } H_2O}{2 \text{ mol } H_2} = 4.89 \text{ mol } H_2O$$

These types of problems are called <u>mole-mole problems</u>, because they require you to calculate moles of one chemical from the given moles of another chemical in the balanced chemical reaction.

You must have a properly-balanced chemical reaction in order to work these types of problems! Furthermore, it must be the **right** reaction. Hydrogen and oxygen can also react to make hydrogen peroxide, H_2O_2:

$$H_2 + O_2 \rightarrow H_2O_2$$

where the reaction stoichiometry is different. Although this reaction is balanced, it is not the correct balanced reaction for the above examples because it does not have water as the product.

Calculations using complicated chemical reactions can be performed just as easily, as the next example shows.

Example 5.5. How many moles of manganese sulfate, $MnSO_4$, will be made as product if 2.89 moles of hydrogen peroxide, H_2O_2, were decomposed according to the following balanced chemical reaction:

$$2\, KMnO_4 + 5\, H_2O_2 + 3\, H_2SO_4 \rightarrow 2\, MnSO_4 + K_2SO_4 + 5\, O_2 + 8\, H_2O$$

Solution. While this is a more complicated chemical reaction, it is balanced, and we can construct proper conversion factors. The chemical reaction says that when 5 moles of H_2O_2 are reacted, 2 moles of $MnSO_4$ are produced. We therefore make the following conversion factor:

$$\frac{2 \text{ mol } MnSO_4}{5 \text{ mol } H_2O_2}$$

and we can calculate the answer:

$$2.89 \text{ mol } H_2O_2 \cdot \frac{2 \text{ mol } MnSO_4}{5 \text{ mol } H_2O_2} = 1.16 \text{ mol } MnSO_4$$

You should note that the numbers in the mole-to-mole conversion factors are considered exact numbers and are not considered when determining the number of significant figures in the final answer.

Mass-Mass Problems

It is a simple step to expand these problems to calculate masses of chemicals. Using the atomic or molecular weights, we can calculate the mass of a product or a reactant if we know its number of moles, and vice versa. These types of problems are called mass-mass problems. What we see, though, is that the mole unit still has a central place even in these kinds of problems.

Given the following balanced chemical reaction,

$$Mg + 2\, HCl \rightarrow MgCl_2 + H_2$$

suppose we want to calculate how many grams of hydrogen gas we can make from 100.0 grams of Mg metal. Because the balanced chemical reaction can be thought of in terms of moles, **not grams**, our first step is to calculate the number of moles of Mg we have. Using the fact that the atomic weight of Mg is 24.3 grams per mole to three significant figures, we have

$$100.0 \text{ g Mg} \cdot \frac{1 \text{ mol Mg}}{24.3 \text{ g Mg}} = 4.12 \text{ mol Mg}$$

where we have started the calculation with what we were given (100.0 g Mg) and have written our conversion factor so that the "g Mg" unit cancels.

Now we can use the balanced chemical reaction to determine how many moles of hydrogen gas is made. For every mole of Mg, one mole of H_2 is given off. Therefore, 1 mol Mg = 1 mol H_2 and we can convert to moles of hydrogen:

$$4.12 \text{ mol Mg} \cdot \frac{1 \text{ mol } H_2}{1 \text{ mol Mg}} = 4.12 \text{ mol } H_2$$

Note how the "mol Mg" units cancel, leaving "mol H_2" as the unit. The question asks for number of grams of hydrogen, so we must perform a third step to convert moles of hydrogen to grams of hydrogen. By adding up the atomic weights of the two hydrogen atoms, we find that one mole of hydrogen has a mass of 2.016 grams. We want to cancel the "mol H_2" unit, so that part of the conversion factor goes into the denominator:

$$4.12 \text{ mol } H_2 \cdot \frac{2.016 \text{ g } H_2}{1 \text{ mol } H_2} = 8.31 \text{ g } H_2$$

So, we can calculate that we get a little over 8 grams of hydrogen gas when we react 100.0 grams of magnesium.

Problems like these can be done in many short steps, like we did above, or in one big step:

$$100.0 \text{ g Mg} \cdot \frac{1 \text{ mol Mg}}{24.3 \text{ g Mg}} \cdot \frac{1 \text{ mol } H_2}{1 \text{ mol Mg}} \cdot \frac{2.016 \text{ g } H_2}{1 \text{ mol } H_2} = 8.30 \text{ g } H_2$$

Notice the existence of truncation error in this problem, too. The answer should be about the same no matter which way you work out the problem. Practice with your calculator to make sure that you get 8.30

as your final answer. You should also check the expressions to see if the appropriate units all cancel correctly, if you work out a problem like this in one long calculation.

Other Types of Problems

The mole-mole and mass-mass problems are a large part of the problems you can work in chemistry. In a sense, they are all the same: you take an initial amount and convert it, using definitions and the balanced chemical equation.

These types of problems can be expanded to include other kinds of conversion factors, like density or molarity or the definition of mole. In fact, in chemistry there will be many equivalencies that can be used as conversion factors in problems such as these. The following examples illustrate some of them. You should be able to do any problem of this type, if given the appropriate information to use as conversion factors.

Example 5.6. How many milliliters of a 0.2515 M solution of potassium hydroxide, KOH, are required to react with 36.22 mL of 0.1889 M sulfuric acid, H_2SO_4? The balanced chemical reaction is

$$2\,KOH + H_2SO_4 \rightarrow 2\,H_2O + K_2SO_4$$

and occurs in aqueous solution.

Solution. Since we ultimately relate everything in terms of moles, we will start by using the molarity unit as a conversion from volume to moles for sulfuric acid. First, we convert the units of volume to liters:

$$36.22 \; \cancel{mL} \cdot \frac{1\,L}{1000 \; \cancel{mL}} = 0.036\,22\,L$$

Now we use the definition of molarity to determine the number of moles of H_2SO_4 that are present:

$$0.1889\,M = \frac{\#\,\text{moles } H_2SO_4}{0.036\,22\,L\;H_2SO_4}$$

$$\#\,\text{moles } H_2SO_4 = (0.1889\,M)\cdot(0.036\,22\,L)$$

$$\#\,\text{moles } H_2SO_4 = 0.006\,842\,\text{mol}$$

Recall that since molarity is defined as $M = \dfrac{mol}{L}$, the product "M·L" equals "mol." Now we convert this number of moles of H_2SO_4 to the number of moles of KOH that will react with it. Using the fact that the balanced chemical reaction states that 2 moles of KOH are needed to react with each mole of H_2SO_4:

$$0.006\,842 \;\cancel{mol\,H_2SO_4} \cdot \frac{2 \text{ mol KOH}}{1 \;\cancel{mol\,H_2SO_4}} = 0.013\,68 \text{ mol KOH}$$

Finally, knowing the molarity of the KOH solution, we can determine the number of mL needed to supply this number of moles of KOH. Again, using the definition of molarity:

$$0.2515 \text{ M KOH} = \frac{0.013\,68 \text{ mol KOH}}{\# \text{ L KOH}}$$

Solving for volume, we cross-multiply and get

$$\# \text{ L KOH} = 0.054\,39 \text{ L} = 54.39 \text{ mL}$$

where we have converted the final answer to milliliters implicitly.

Example 5.7. The density of dry air is approximately $1.057 \;\dfrac{g}{L}$ at 0°C and 1 atm pressure. Assuming that the average molecular weight of air is 28.8 grams per mole, estimate the number of gas molecules per cubic centimeter in air. Recall that there are 6.02×10^{23} molecules in a mole of molecules.

Solution. What we need is a conversion factor based on the following relationship:

$$1 \text{ mole air} = 28.8 \text{ grams of air}$$

We will also use the fact that $1 \text{ L} = 1000 \text{ cm}^3$. Starting with what we are given, the density, we first convert into units of moles per liter:

$$1.057 \frac{\cancel{g}}{L} \cdot \frac{1 \text{ mol}}{28.8 \;\cancel{g}} = 0.0367 \frac{mol}{L}$$

Now we convert to molecules per liter:

$$0.0367\,\frac{mol}{L} \cdot \frac{6.02 \times 10^{23}\text{ molecules}}{1\,mol} = 2.21 \times 10^{22}\,\frac{\text{molecules}}{L}$$

Finally, we convert the liters to cubic centimeters, getting

$$2.21 \times 10^{22}\,\frac{\text{molecules}}{L} \cdot \frac{1\,L}{1000\text{ cm}^3} = 2.21 \times 10^{19}\,\frac{\text{molecules}}{\text{cm}^3}$$

as our final answer. We could have performed all conversions in one long line:

$$1.057\,\frac{g}{L} \cdot \frac{1\text{ mol}}{28.8\text{ g}} \cdot \frac{6.02 \times 10^{23}\text{ molecules}}{1\text{ mol}} \cdot \frac{1\,L}{1000\text{ cm}^3} = 2.21 \times 10^{19}\,\frac{\text{molecules}}{\text{cm}^3}$$

The numerical answer would be the same. You should convince yourself that the units do cancel out properly!

Student Exercises

5.1. Balance the following reactions.

(a) $NaOH + H_2SO_4 \rightarrow Na_2SO_4 + H_2O$

(b) $BaCl_2 + Al_2(SO_4)_3 \rightarrow BaSO_4 + AlCl_3$

(c) $NaHCO_3 + HCl \rightarrow NaCl + H_2O + CO_2$

5.2. (a) How many moles of O atoms are there in 1.5 moles of glucose, $C_6H_{12}O_6$? (b) How many moles of H atoms are there in 0.50 moles of sucrose, $C_{12}H_{22}O_{11}$? (c) How many moles of total atoms are there in 2.0 moles of ethanol, C_2H_5OH?

5.3. (a) How many grams are there in 2.50 moles of NaOH? (b) How many moles are there in 100.0 grams of CO_2? (c) How many grams are there in 5.22×10^2 L of argon gas if each mole occupies 24.4 liters of volume? (d) How many moles are present in 25.9 mL of hexane, C_6H_{14}, if hexane has a density of 0.6603 $\dfrac{g}{mL}$?

5.4. Which contains more total atoms, one mole of $Al_2(CO_3)_3$ or two moles of $BaSO_4$?

5.5. Which contains more total atoms, 1.00×10^2 grams of NaCl or 1.00×10^2 grams of LiBr?

5.6. How many moles of water are given off by the metabolism of 1 mole of sucrose? The unbalanced chemical reaction is

$$C_{12}H_{22}O_{11} + O_2 \rightarrow CO_2 + H_2O$$

5.7. How many moles of carbon dioxide are given off by 5.00 moles of sodium bicarbonate which, when heated, decomposes according to the following balanced chemical reaction (which is why $NaHCO_3$ is sometimes used as a fire extinguisher):

$$2\,NaHCO_3 + heat \rightarrow Na_2CO_3 + H_2O + CO_2$$

5.8. If each mole of gas has a volume of 22.4 L at standard temperature and pressure, what volume of CO_2 is given off by the decomposition of 1.00×10^2 g of sodium bicarbonate? See the above problem for the balanced chemical reaction.

5.9. Ammonia is produced industrially by the following gas-phase reaction:

$$N_2 + 3\,H_2 \rightarrow 2\,NH_3$$

(a) How many grams of ammonia will be produced by the complete reaction of 150. (1.50×10^2) grams of hydrogen gas? (b) How many grams of nitrogen are necessary to react with 150. (1.50×10^2) grams of hydrogen? (c) Do these masses satisfy the Law of Conservation of Mass?

5.10. How many milliliters of 0.1107 M HCl solution are needed to react with 1.022 grams of $Ba(OH)_2$? The unbalanced chemical reaction is

$$HCl + Ba(OH)_2 \rightarrow BaCl_2 + H_2O$$

5.11. If the density of 1.450 M HCl solution is 1.065 $\frac{g}{mL}$, what mass of HCl solution is needed to react with 45.02 mL of 2.070 M Ca(OH)$_2$? The balanced chemical reaction is

$$2\ HCl\ +\ Ca(OH)_2\ \rightarrow\ CaCl_2\ +\ 2\ H_2O$$

5.12. The atmosphere on the planet Mars consists almost solely of CO$_2$ and has approximately 6.07 x 10^{16} molecules per cubic centimeter. Calculate the density of Mars' atmosphere.

Answers to Student Exercises

5.1. (a) $2\,NaOH + H_2SO_4 \rightarrow Na_2SO_4 + 2\,H_2O$

(b) $3\,BaCl_2 + Al_2(SO_4)_3 \rightarrow 3\,BaSO_4 + 2\,AlCl_3$

(c) $NaHCO_3 + HCl \rightarrow NaCl + H_2O + CO_2$ (The reaction is already balanced.)

5.2. (a) There are 9.0 moles of O atoms. (b) There are 11 moles of H atoms. (c) There are 18 moles of total atoms.

5.3. (a) There are 100. (1.00×10^2) grams of NaOH. (b) There are 2.273 moles of CO_2. (c) There are 855 grams of Ar gas. (d) There are 0.198 moles of hexane.

5.4. One mole of $Al_2(CO_3)_3$ contains more moles of atoms (14) than do 2 moles of $BaSO_4$ (12).

5.5. 100. grams of NaCl contains 2.06×10^{24} atoms, while 100. grams of LiBr contains 1.39×10^{24} atoms. The 100. grams of NaCl has more total atoms.

5.6. You should get 11 moles of water for every mole of sucrose metabolized.

5.7. You will get 2.50 moles of CO_2 for every 5.00 moles of $NaHCO_3$ that decomposes.

5.8. 13.3 L of carbon dioxide.

5.9. (a) 844 grams of ammonia (b) 694 grams of N_2. (c) Yes, because the total mass of reactants (694 + 150) equals the total mass of products (844).

5.10. We will need 107.8 mL of HCl solution.

5.11. We will need 136.9 grams of HCl solution.

5.12. The density of the Martian atmosphere is 4.44×10^{-3} grams per cubic centimeter, or $0.004\,44\ \dfrac{g}{cm^3}$.

This compares to a density of about $1.05\ \dfrac{g}{cm^3}$ for Earth!

Chapter 6. Using Mathematical Formulas

Introduction

Another important type of math problem that you will need to be able to do in working chemistry problems is one in which a specific equation or formula must be used. Many of these formulas have names: Boyle's law, Charles' law, the ideal gas law, the Nernst equation. Many of them don't have names. Most of them will probably have to be memorized. But even if the equations are known (whether they are given to you or if you know them by memory), you still need to put quantities into the equation properly and to evaluate the unknown quantity correctly. Although we reviewed some of the algebra necessary to solve for unknowns in equations in Chapter 2, in this chapter we will focus more on formulas and how to use them.

Formulas

In the mathematical sense, a _formula_ is an equation that relates different quantities to each other. (Chemistry also uses _chemical formulas_ to indicate the compositions of elements and compounds, but we are not using the word "formula" in that way here.) Many of the definitions we have used in previous chapters are in fact very simple formulas. The definition of molarity,

$$\text{molarity} = \frac{\text{moles of solute}}{\text{liters of solution}}$$

can be considered a formula. It relates a concentration, in units of molarity, to the number of moles of solute and the volume of the solution in liters.

Most of the mathematical formulas we use in chemistry either relate seemingly non-related quantities to each other or allow us to determine how a quantity changes with other changes in conditions. For example, the ideal gas law is

$$PV = nRT$$

and relates the pressure (P), volume (V), number of moles (n), and the temperature (T) of a sample of gas. R is called the _ideal gas law constant_ and is the proportionality constant that relates all of these quantities of the gas. Most formulas are expressed in terms of variables that represent various quantities and make the equation easier to write. The ideal gas law could be written as "(pressure)(volume) = (number of moles)(ideal gas law constant)(temperature)", but "$PV = nRT$" is more compact. In this case, the formula relates **instantaneous** conditions, or conditions of the gas as it exists right now. As an example of a formula that relates a change in conditions, consider Boyle's law,

$$P_1V_1 = P_2V_2$$

which relates an initial pressure and volume P_1 and V_1 with the values of some final pressure and volume P_2 and V_2, which have both changed in value. (Boyle's law presumes that the both temperature and amount of the gas stay constant as the pressure and volume change.) Formulas like this relate conditions of a system that change and show that some of the measurable quantities that describe a system change together, in a certain relationship.

Most formulas simply require that we substitute values for each of the variables in the equation except for one, and then algebraically solve for that one unknown variable. "That's easy," many students think. But many students don't do it **properly**. They get the wrong answer, and don't understand why. Although we will not be able to cover every possible formula you might encounter, hopefully the ideas that this chapter covers will help you properly use any formula to determine the correct desired quantity.

Plugging In Properly

In virtually all cases when working with a formula, you will know – or can find – the value for every variable in the equation except for one. This is usually the variable you will be asked to determine. What you need to be able to do is to properly substitute values for the variables, then perform some algebra to solve for the unknown quantity.

The first step to being able to use a formula properly is so basic that most people don't even recognize its importance: **know what the variables stand for.** In the ideal gas law equation given above, we defined what P, V, n, R, and T were. When you know what the variables stand for, you can substitute the proper numbers and units in the correct places in the formula. If you do not know what the variables stand for, there would be no way you could properly use the formula to determine some unknown quantity.

Why is this important to understand? Because science deals with a lot of different quantities, each of which has its own specific variable. Some of these variables may not be obvious, either. Using "P" for pressure and "V" for volume may be obvious, but using "R" for the ideal gas law constant is not. (The letter "r" doesn't even appear in the phrase "ideal gas law constant"!) Furthermore, because of the vast number of quantities that science defines, we are forced to use the same variable for different quantities.

R, for example, is used to symbolize the ideal gas law constant, the Rydberg constant (used to discuss electronic structure of atoms), and roentgens (a unit of radiation exposure). **It is crucial that you know what the variables stand for in any formula you use.**

When you replace a variable by a specific quantity, you do not just substitute the number from the quantity. **You must substitute the unit of that quantity also.** We will see shortly how important units are in working with formulas.

Example 6.1. Given that a 0.773 mol sample of a gas has a temperature of 298 K, a volume of 12.7 L, and a pressure of 1.49 atm, rewrite the ideal gas law in terms of these values. The constant R has a value of 0.082 05 $\dfrac{\text{L} \cdot \text{atm}}{\text{mol} \cdot \text{K}}$.

Solution. The pressure and volume are multiplied together on one side of the equation, whereas the amount, temperature, and R are multiplied together on the other side of the equation. We get

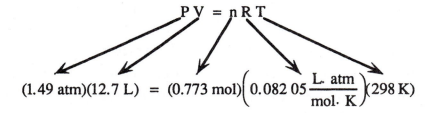

$$P V = n R T$$

$$(1.49\ \text{atm})(12.7\ \text{L}) = (0.773\ \text{mol})\left(0.082\ 05\ \frac{\text{L} \cdot \text{atm}}{\text{mol} \cdot \text{K}}\right)(298\ \text{K})$$

where the arrows show the correct values substituted for the variables. Keep in mind that because multiplication is commutative, you don't have to write the values in the exact order as the variables are in the formula. You just have to multiply pressure and volume on one side and equate that product to the amount, the temperature, and the R values.

While the above example may have seemed trivial, there is another point to be made. When plugging into formulas, **the units must be consistent in all parts of the formula**. Let us take a closer look at the expression above. Units will cancel algebraically in two different ways: first, if the same unit is in both the numerator and denominator on the same side of an equation, and second, if the same unit were on both sides in the numerator or on both sides in the denominator of the equation.

Let us see how the units work in the formula in Example 6.1. On the right side, the units for temperature (degrees Kelvin) and amount (moles) cancel because they are in the numerator and denominator on the same side of the equation:

$$(1.49 \text{ atm})(12.7 \text{ L}) = (0.773 \,\cancel{\text{mol}})\left(0.082\,05\,\frac{\text{L·atm}}{\cancel{\text{mol·K}}}\right)(298\,\cancel{\text{K}})$$

Next, we notice that the units "liters" and "atmospheres" are in the numerator on both sides of the equation. Therefore, they cancel algebraically:

$$(1.49 \,\cancel{\text{atm}})\,(12.7 \,\cancel{\text{L}}) = (0.773)(0.082\,05\,\cancel{\text{L}}\,\cancel{\text{atm}})\,(298)$$

(Here, we have omitted the mole and temperature units because we canceled them in the previous step.) What we have left are simply some numbers that we need to multiply together:

$$(1.49)(12.7) = (0.773)(0.082\,05)(298)$$

$$18.9 = 18.9$$

The final equation is what is called a "reflexive relationship" in algebra and is understood to be true (i.e. something equals itself). The point here is that the units of each type of quantity (volume, pressure, temperature, amount) are all consistent with each other. Units must be consistent when plugging values into a formula.

Let us consider an example where the units are **not** consistent with each other. Again, we will use the ideal gas law. If we are given a 0.001 90 mol sample of gas that has a volume of 55.8 mL, a pressure of 658 torr, and a temperature of 37.0°C, we **could** plug into the ideal gas law directly and get

$$(658 \text{ torr})(55.8 \text{ mL}) = (0.00190 \text{ mol})\left(0.082\,05\,\frac{\text{L·atm}}{\text{mol·K}}\right)(37°\text{C})$$

The problem with doing this is that the units will not cancel! We are using two different units for pressure, two different units for volume, and two different units for temperature. Only the "mole" unit will cancel properly.

This is not a permanent problem, **since we can convert units** and make them consistent with each other. For most types of variables, it does not matter which unit we convert to as long as all variables of the same type (volume, pressure, energy, etc.) are expressed in the same units. The exception is temperature, which in most formulas should always be expressed in Kelvin units.

We can convert all pressure units to atmospheres (there are exactly 760 torr in 1 atm), and all volume units to liters, and get

$$(0.866 \text{ atm})(0.0558 \text{ L}) = (0.001\ 90 \text{ mol})\left(0.082\ 05\ \frac{\text{L} \cdot \text{atm}}{\text{mol} \cdot \text{K}}\right)(310 \text{ K})$$

(where we have also converted temperature into Kelvin units). **Now** the formula is correctly set up, and the units cancel and the numbers combine to get

$$0.0483 = 0.0483$$

which we know is true.

Example 6.2. Instead of converting pressure and volume units to atm and L, convert all pressure and volume units to torr and mL and show that the formula is still an equality.

Solution. Only the value for R has units that need to be converted. We use the relationship between L and mL, and the fact that 760 torr = 1 atm:

$$0.082\ 05\ \frac{\cancel{\text{L}} \cdot \cancel{\text{atm}}}{\text{mol} \cdot \text{K}} \cdot \frac{760 \text{ torr}}{1 \cancel{\text{atm}}} \cdot \frac{1000 \text{ mL}}{1 \cancel{\text{L}}} = 62,360\ \frac{\text{mL} \cdot \text{torr}}{\text{mol} \cdot \text{K}}$$

Using this value for R (it's the same R, just different units), we get

$$(658 \text{ torr})(55.8 \text{ mL}) = (0.001\ 90 \text{ mol})\left(62,360\ \frac{\text{mL} \cdot \text{torr}}{\text{mol} \cdot \text{K}}\right)(310 \text{ K})$$

The units all cancel, and we are left with

$$(659)(55.8) = (0.001\ 90)(62,360)(310)$$

$$36,700 = 36,700$$

which is still an equality. The number is different, but that's only because we are using different units.

The point is that it doesn't matter which units you convert, as long as the units are the same. Otherwise, you will not be able to perform the correct algebra with the units.

Using Formulas to Solve for Unknowns

Most of the time, we use formulas to solve for some unknown quantity. Typically, we know all but one of the variables – either they're given in the problem or we can look up data in tables – and we need to calculate that unknown variable.

First, of course, you need to know what formula to use to solve a problem. Next, you need to plug in all the information you have. You will need to make sure that all of the units are consistent, and that they will cancel so that the only unit left is an appropriate unit for the quantity you are seeking. Finally, you evaluate all of the numbers to calculate your final answer. Keep in mind that you may have to algebraically manipulate the formula so that the variable you are looking for is by itself in the numerator.

A simple equation might be

$$\Delta G° = -n \mathcal{F} E°$$

where $\Delta G°$ is the change in the Gibbs free energy of an oxidation-reduction ("redox") reaction, n is the number of moles of electrons transferred in the reaction, $E°$ is the voltage (symbol = V) of the redox reaction, and $\mathcal{F}$ is the <u>Faraday constant</u> and equals 96,500 coulombs (symbol = C) per mole of electrons. If we are given that the number of moles of electrons transferred is 2, and that the voltage of the redox reaction is +1.10 volts, then we simply plug in to get

$$\Delta G° = -(2 \text{ moles of electrons}) \left(96,500 \frac{C}{\text{mole of electrons}} \right) (1.10 \text{ V})$$

The "moles of electrons" units cancel and we multiply the numbers to get

$$\Delta G° = -212,000 \text{ C·V}$$

These units look funny until we learn that a coulomb-volt is equal to a joule, a unit of energy. Therefore, we write

$$\Delta G° = -212,000 \text{ J} = -212 \text{ kJ}$$

If we are given that the $\Delta G°$ for a three-electron reaction is +23.07 kJ and want to calculate the $E°$ for the process, we will need to do some algebraic rearranging of the same formula. The rearranging can be done before or after the values are plugged into the formula. If the algebra is done correctly, it should not matter which way you do it.

For example, we can take the formula

$$\Delta G° = -n\mathcal{F}E°$$

and divide both sides by -n and $\mathcal{F}$ to get

$$E° = -\frac{\Delta G°}{n\mathcal{F}}$$

and now plug in our known values, converting our $\Delta G°$ into units of J:

$$E° = -\frac{23{,}070 \text{ J}}{(2 \text{ mol e}^-)\left(96{,}500\dfrac{\text{C}}{\text{mol e}^-}\right)} = -0.120\frac{\text{J}}{\text{C}} = -0.120 \text{ V}$$

Or, you can plug into the original equation, cancel the units, and multiply and divide the numbers appropriately to isolate $E°$ all by itself in the numerator:

$$23{,}070 \text{ J} = -(2 \text{ mol of electrons})\left(96{,}500\dfrac{\text{C}}{\text{mol of electrons}}\right)(E°)$$

$$23{,}070 \text{ J} = -193{,}000 \text{ C} \cdot E°$$

$$E° = -\frac{23{,}070 \text{ J}}{193{,}000 \text{ C}} = -0.120\frac{\text{J}}{\text{C}} = -0.120 \text{ V}$$

Either way, we get the same answer, **as long as the algebra is performed properly.** If you are uncertain about your algebra ability, it might be easier to plug the numbers into the original equation first, then cancel units and do algebra with the numbers, rather than the variables.

Example 6.3. The combined gas law relates the pressure, volume, and absolute temperature of a gas and how they change with respect to each other. If P_1, V_1, and T_1 were the initial conditions of the gas and they changed to some final conditions P_2, V_2, and T_2, the combined gas law requires that

$$\frac{P_1 V_1}{T_1} = \frac{P_2 V_2}{T_2}$$

If the initial conditions were 1.50 L, 874 torr, and 295 K, and the final conditions were 2.95 L and 1.78 atm, what is the final temperature?

Solution. First, we should take the time to recognize what values are given to us. From the units, we can determine that the initial volume V_1 = 1.50 L, the initial pressure P_1 = 874 torr, and the initial temperature T_1 = 295 K. The final volume V_2 = 2.95 L, the final pressure P_2 = 1.78 atm, and the final temperature T_2 is the unknown. Notice that the pressures are given in two different units, and we must convert one of the values so that both pressure values have the same units. Let us convert 1.78 atm to units of torr. Since 760 torr = 1 atm:

$$1.78 \text{ atm} \cdot \frac{760 \text{ torr}}{1 \text{ atm}} = 1350 \text{ torr}$$

We will use 1350 torr as our final pressure. Note that we could have converted the torr unit to atmospheres, but either way will give us the same final answer if we do all of the algebra correctly.

There are several ways to approach the formula in this example. Perhaps the easiest will be to cross-multiply the two fractions,

$$\frac{P_1 V_1}{T_1} \diagdown\!\!\!\!\times\!\!\!\!\diagup \frac{P_2 V_2}{T_2}$$

$$T_2 P_1 V_1 = T_1 P_2 V_2$$

and then divide both sides by $P_1 V_1$:

$$\frac{T_2 \cancel{P_1 V_1}}{\cancel{P_1 V_1}} = \frac{T_1 P_2 V_2}{P_1 V_1}$$

This isolates T_2 in the numerator on one side of the equation, all by itself:

$$T_2 = \frac{T_1 P_2 V_2}{P_1 V_1}$$

Plugging in all of our known values, we get

$$T_2 = \frac{(295\,K)(1350\,\cancel{torr})(2.95\,\cancel{L})}{(874\,\cancel{torr})(1.50\,\cancel{L})}$$

Notice that all of the units cancel except for K, a unit of temperature, which is what we're looking for. The final answer, after multiplying and dividing all of the numbers, is

$$T_2 = 896\,K$$

You should verify this by entering all of these numbers into your calculator to see if you can get the same answer.

Again, it is important to keep track of the units: the units of your final answer **must** be an appropriate unit for the quantity you are calculating. In the above example, we found units of degrees Kelvin, which is the proper unit for temperature. If you are solving for a pressure, the final unit should be atm, torr, mmHg, bar, or some other unit of pressure. If you are calculating an energy, the final unit had better be calories, joules, kilojoules, etc. These are appropriate units of energy. If you do not get an appropriate unit, **then there is probably something wrong with your calculation.** In time, you will find that keeping track of your units is a tool you can use to help you keep track of the progress of your calculation.

Formulas can get more complicated. Consider the following example:

$$\Delta G = \Delta G° + RT \ln Q$$

which allows you to calculate the change in the Gibbs free energy at any condition, ΔG, from the ΔG at standard conditions, $\Delta G°$ (notice the superscript °), and a correction given by $RT \ln Q$. The "ln" refers to the natural logarithm, which we will look at in the next chapter. Q is called the reaction quotient. It is the "products-over-reactants" expression and depends on the specific chemical reaction of interest. (For additional information about reaction quotients, consult your chemistry textbook.) For instance, for the ionization reaction

$$CH_3COOH \rightleftharpoons CH_3COO^- + H^+$$

the expression for Q is

$$Q = \frac{[H^+][CH_3COO^-]}{[CH_3COOH]}$$

The variables in brackets refer to concentrations of each chemical species. (For a more detailed discussion of Q and how it relates to reactions, consult your chemistry textbook.) So, for that chemical reaction – and **only** for that chemical reaction – the complete formula to calculate ΔG is

$$\Delta G = \Delta G^\circ + RT \ln \frac{[H^+][CH_3COO^-]}{[CH_3COOH]}$$

Although this formula is a little more complicated than the ones we have discussed previously, the same ideas apply: you should be able to solve for any one of the variables in the formula – even the individual concentrations of the chemical products and reactants.

This equation for ΔG brings up another issue concerning units. As an energy change, ΔG has units of joules or joules per mole. This means that the second term on the right, $RT \ln \frac{[H^+][CH_3COO^-]}{[CH_3COOH]}$, also must have units of joules or joules per mole. A logarithm is a pure number and doesn't have units, but R and T do. We usually have used R as equal to $0.082\ 05\ \frac{L \cdot atm}{mol \cdot K}$, and with temperature having units of K, the combination of the two variables has the overall unit of $\frac{L \cdot atm}{mol}$. This is not a typical unit of energy, and **not** equal to $\frac{J}{mol}$. However, we have seen that we can express R with different units. It will have a different numerical value, but that's OK. There is a value for R that has joule units: $R = 8.314\ \frac{J}{mol \cdot K}$. When we use this value for R, the units work out properly for calculating a change in energy. Again, we stress the point that the units of the variables you plug into formulas must be consistent.

Example 6.4. Given the chemical reaction

$$N_2 + 3\,H_2 \rightleftharpoons 2\,NH_3$$

if you were given that ln Q = 2.568 at a temperature of 450°C and ΔG = -1.26 kJ/mol, calculate $\Delta G°$ for the reaction.

Solution. Using the formula

$$\Delta G = \Delta G° + RT \ln Q$$

we should recognize that the problem gives us ΔG, T, ln Q, and we know a value for R. We therefore plug in for everything but one variable, which is the one we are looking for: $\Delta G°$.

$$-1.26 \text{ kJ/mol} = \Delta G° + \left(8.314 \, \frac{J}{mol \cdot K} \right)(450 + 273 \text{ K}) \, (2.568)$$

where we have converted our temperature into degrees Kelvin. The temperature units on the right side cancel and we have

$$-1.26 \text{ kJ/mol} = \Delta G° + 15,400 \text{ J/mol}$$

In order to complete the problem, we need to convert units so that they are the same for all terms. Either the J/mol can be converted to kJ/mol, or the kJ/mol can be converted to J/mol. Let us convert the units on the last term to kJ/mol:

$$-1.26 \text{ kJ/mol} = \Delta G° + 15.4 \text{ kJ/mol}$$

Only when the units are the same can the numbers be combined. We subtract 15.4 kJ/mol from both sides in order to isolate the quantity we are solving for:

$$-16.7 \text{ kJ/mol} = \Delta G°$$

as our final answer.

In all of the above examples, we have been using many of the topics of previous chapters. We have limited our final answer to the tenths' place, according to the rules of significant figures. Notice, too, that in working this example, we followed the proper order of operations that were discussed in Chapter 2. The proper math skills must already be developed if a formula is going to be used to determine the correct answer.

Student Exercises

In many of the following exercises, you will be given a formula and some of the quantities, and you will then be asked to solve for one of the variables. Watch your units!

6.1. Another gas law, called <u>Amontons' law,</u> is given by the mathematical formula

$$\frac{V_1}{T_1} = \frac{V_2}{T_2}$$

In words, identify the variables. (HINT: see Example 6.3.)

6.2. Write a formula that expresses the following statement mathematically: "The force of gravity between two masses is equal to a proportionality constant times the product of the two masses and divided by the square of the distance between the masses."

6.3. Rewrite the ideal gas law to solve for R, the ideal gas law constant.

6.4. Use the equation $c = \lambda \cdot v$, where c is the speed of light (which is constant and equal to 2.9979×10^8 $\frac{m}{sec}$), λ is the wavelength of the light in units of meters, and v is the frequency in units of $\frac{1}{sec}$, to calculate the frequency of light that has a wavelength of 5.50×10^{-7} m. Do you see how the units work out?

6.5. Use the equation $E = h \cdot v$, where E is the energy of a single photon, v is the frequency in units of $\frac{1}{sec}$, and h is Planck's constant and is equal to 6.626×10^{-34} J·sec (joule-seconds), to calculate the energy of a single photon that has a wavelength of 5.50×10^{-7} m. See Exercise 6.4 for how to calculate the frequency v of the photon.

6.6. A spectrum is the absorption or emission of light of specific wavelengths by chemicals. The spectrum of hydrogen atoms is particularly simple because hydrogen is the simplest element. In fact, in 1885 Balmer showed that the wavelengths of visible light in the hydrogen spectrum could be predicted by the simple formula

$$\frac{1}{\lambda} = 10,973,700 \left(\frac{1}{4} - \frac{1}{n^2} \right) \frac{1}{m}$$

where the unit for $\frac{1}{\lambda}$ is $\frac{1}{m}$. The variable n is an integer greater than 2: n = 3, 4, 5, etc. Calculate to three significant figures the wavelength λ of the line in the hydrogen spectrum corresponding to n = 5.

6.7. Use the combined gas law to determine the final pressure for a gas that has initial conditions 345 mL, 1.007 atm, and 25.0°C and final conditions 100°C and 1.28 L.

6.8. Use the ideal gas law to calculate the number of moles of gas that occupy a volume of 1.00 liters with a pressure of 1.0×10^3 torr and a temperature of 37.0°C. The additional information you need to solve this problem is located within the chapter text.

6.9. Use the equation $\Delta G° = -n\mathcal{F}E°$ to calculate $\mathcal{F}$ for a redox reaction that involves two moles of electrons and has a voltage of 1.107 V and a $\Delta G°$ of -212.3 kJ.

6.10. The <u>Nernst equation</u> is a formula for calculating the voltage of a battery at non-standard conditions:

$$E = E° - \frac{RT}{n\mathcal{F}} \ln Q$$

where E is the non-standard voltage, E° is the voltage of the battery under standard conditions, and the rest of the variables have their normal meaning. (a) Identify the other variables R, T, n, $\mathcal{F}$, and ln Q. (b) If the voltage of a battery is +0.453 V at a temperature of 37°C for a redox reaction that involves 5 moles of electrons and has ln Q = -3.775, what is the voltage of the battery at standard conditions? (Note that the formula doesn't even require that you know what "standard conditions" are!) You will need to use $\mathcal{F} =$ 96,500 C/mol e⁻ and $R = 8.314 \dfrac{J}{mol \cdot K}$.

6.11. In 1913, scientist Neils Bohr derived a formula for the <u>Rydberg constant,</u> R, which is part of the formula that describes the hydrogen atom spectrum. (See Exercise 6.6 above.) Bohr found that

$$R = \frac{m_e e^4}{8\varepsilon_0^2 h^2}$$

where the constants in the formula and their values are:

Symbol	Meaning	Value
m_e	mass of electron	9.109×10^{-31} kg
e	charge on electron	1.602×10^{-19} C
ε_0	permittivity of free space	$8.854 \times 10^{-12} \dfrac{C^2}{J \cdot m}$
h	Planck's constant	6.626×10^{-34} J·sec

(a) Substitute the values into the formula and calculate a value for R. You may have to be careful if your calculator does not allow three-digit powers of ten. What are the final units you get? (HINT: Remember that joule is a derived unit. What is its definition in terms of fundamental units?)

(b) The value of R in Exercise 6.6 is $10,973,700 \dfrac{1}{m}$. Using the equations $E = h\nu$ and $c = \lambda\nu$, can you convert your answer in (a) to this value and this unit?

Answers to Student Exercises

6.1. V_1 is the initial volume, T_1 is the initial temperature, V_2 is the final volume, and T_2 is the final temperature.

6.2. Your formula might look something like this: $F = k \cdot \dfrac{m_1 \cdot m_2}{r^2}$, where F is the force of gravity, k is the proportionality constant, m_1 and m_2 are the masses, and r is the distance between the masses. You might have used different symbols for the variables, but the form of your equation should be similar.

6.3. $R = \dfrac{PV}{nT}$

6.4. $v = 5.45 \times 10^{14} \ \dfrac{1}{sec}$, or 5.45×10^{14} sec^{-1}.

6.5. $E = 3.61 \times 10^{-19}$ J for a single photon.

6.6. $\lambda = 4.34 \times 10^{-7}$ m for the spectral line corresponding to n = 5. In case you're curious, that wavelength corresponds to light having a violet color, almost out of range of visible light.

6.7. The final pressure is 0.340 atm.

6.8. $n = 5.2 \times 10^{-2}$ mol, or 0.052 mol of gas.

6.9. Using the data given, we calculate a value for $\mathcal{F}$ as 95,890 C/mol e$^-$.

6.10. (a) R is the ideal gas law constant, T is the absolute temperature, n is the number of electrons transferred in the redox reaction, $\mathcal{F}$ is the Faraday constant, and ln Q is the natural logarithm of the reaction quotient, the normal "products-over-reactants" expression for a chemical reaction. (b) $E^\circ = 0.473$ V.

6.11. (a) Simply plugging into the expression, you should get 2.179×10^{-18} J. (b) Using the two equations, you should be able to get that $R = 10,970,000 \ \dfrac{1}{m}$, which to four significant figures is the same value for R that was used in the previous exercise.

Chapter 7. Advanced Math Topics

Introduction

Don't let the chapter title scare you! It's just that occasionally in general chemistry, there are some specialized math tools that must be used in order to work out a problem. Such topics include things like the quadratic formula, logarithms, exponentials, and roots. Since many of these topics are math skills that are only used in the latter part of a general chemistry sequence, we have put off discussing them until now.

Exponents

Most of the equations we have used so far have been composed of different variables multiplied, divided, added to, and subtracted from each other. Usually, the variable occurs once in the equation. Since it is understood that these variables have an exponent of one, we state that the variables are "raised to the first power." Variables that are raised to the first power are the most common variables we encounter, and one is the presumed exponent if no exponent is written explicitly.

When we multiply a variable by itself, like $V \times V$, we write it as V^2 and say that the variable is "raised to the second power" or that the variable is "squared." If three of the same variable are multiplied, like $V \times V \times V$, we write that as V^3, which is the variable "raised to the third power," or "cubed." Similarly, we have V^4 as "fourth power," V^5 as "fifth power," etc. (There are no common alternate names for variables raised to any power other than 2 or 3.)

If two different powers of the same variable are multiplied together, the result will be the variable raised to a power equal to the sum of the exponents. For example,

$$V^2 \cdot V^3 = V^{(2+3)} = V^5$$

This rule works with numbers as well as variables. The most common occurrence will be with powers of ten; for example,

$$10^5 \cdot 10^8 = 10^{(5+8)} = 10^{13}$$

For division of the same variable raised to different exponents, the result will be the variable raised to a power equal to the exponent of the numerator minus the exponent in the denominator. For example,

$$\frac{V^5}{V^2} = V^{(5-2)} = V^3$$

Negative numbers can also be used as exponents. A negative exponent implies the reciprocal of the variable raised to the positive exponent. For example,

$$V^{-1} \text{ means } \frac{1}{V^1} = \frac{1}{V}$$

In a similar fashion,

$$V^{-3} \text{ means } \frac{1}{V^3}$$

and so forth. If a variable with a negative exponent is in the denominator of a fraction, its reciprocal places it in the **numerator** with a positive exponent:

$$\frac{1}{V^{-3}} = \frac{1}{\frac{1}{V^3}} = 1 \div \frac{1}{V^3} = 1 \cdot \frac{V^3}{1} = V^3$$

In the next to last step, we have used the fact that division is equal to multiplication by the reciprocal. (See Chapter 2.)

Example 7.1. Simplify the following expression so that all variables have positive exponents.

$$\frac{x^{-3}y^2}{w^{-1}z^{-4}}$$

Solution. In order to see the individual changes better, we will separate the four variables in the expression so that every variable in the numerator is written as a fraction over one and the variables in the denominator are written as one divided by the variable.

$$\frac{x^{-3}y^2}{w^{-1}z^{-4}} = \frac{x^{-3}}{1} \cdot \frac{y^2}{1} \cdot \frac{1}{w^{-1}} \cdot \frac{1}{z^{-4}}$$

(You should verify that this is correct.) The negative exponents on x, w, and z mean that we can rewrite those variables as the reciprocal raised to the positive power. Rewriting:

$$\frac{x^{-3}}{1} \cdot \frac{y^2}{1} \cdot \frac{1}{w^{-1}} \cdot \frac{1}{z^{-4}} = \frac{1}{x^3} \cdot \frac{y^2}{1} \cdot \frac{w^1}{1} \cdot \frac{z^4}{1}$$

Notice that the y^2 term did not change. Recombining the four terms into a single fraction, we get

$$\frac{y^2 w^1 z^4}{x^3}$$

as our final answer with all positive exponents. Normally, the "1" exponent on the w variable is not written explicitly, but it is included here to illustrate the principle.

Variables are not the only things that can have negative exponents; units can as well. This might seem unusual unless you remember that units can also be in denominators of fractions and can be rewritten in a numerator if given a negative exponent. In fact, we have already worked with a unit that can be expressed with a negative exponent. When we consider the frequency of a wave (of light, for example), we speak of the "number of waves per second" that pass a particular point. The "number of waves" is simply a number, but the "per second" is a unit. Since frequency has no unit in the numerator, only in the denominator, frequency has units of $\frac{1}{sec}$. Using negative exponents, this unit is sometimes written as sec^{-1} or s^{-1}. (Science defines the unit "hertz" as equal to $\frac{1}{sec}$. The hertz unit has the abbreviation "Hz" and is named after Heinrich Hertz, who discovered radio waves.)

Derived units can also be expressed using negative exponents. For example, the unit $\frac{kJ}{mol}$ can be rewritten as

$$kJ \cdot \frac{1}{mol} = kJ \cdot mol^{-1}$$

You might occasionally see "$kJ \cdot mol^{-1}$" instead of "$\frac{kJ}{mol}$," but you should recognize that the two expressions mean the same thing.

Example 7.2. Write the fundamental units for the newton so that all units are in the numerator, with the appropriate units having negative exponents.

Solution. Recall the definition of the newton:

$$N = \frac{kg \cdot m^2}{sec^2}$$

We will rewrite the fraction like we did in Example 7.1:

$$\frac{kg \cdot m^2}{sec^2} = \frac{kg}{1} \cdot \frac{m^2}{1} \cdot \frac{1}{sec^2}$$

$$= \frac{kg}{1} \cdot \frac{m^2}{1} \cdot \frac{sec^{-2}}{1}$$

$$= kg \cdot m^2 \cdot sec^{-2}$$

You may think that this is an extreme example. However, many books, articles, and reference materials express complicated units for quantities and constants in this manner. Can you find any examples in your own textbook?

The Quadratic Formula

In certain problems dealing with equilibrium concentrations, an equation arises that has an unknown – usually called "x" – raised to the second power. There may or may not be another term in the equation with the variable x raised to the first power, and there may or may not be a constant as part of the equation. The equation is equal to zero. For example, one such equation may look like this:

$$x^2 - 2x + 1 = 0$$

Any equation that is the combination of terms where the highest exponent on the variable is 2 is called a quadratic equation. A quadratic equation can be thought of as a combination of three powers of x: x raised to the second power, x raised to the first power, and x raised to the zero power. (Recall that anything raised to the zero power is one.) A proper quadratic equation equals zero, which may require that an equation be algebraically rewritten. The numbers that precede each power of x are called <u>coefficients</u>. In the above quadratic equation, the coefficient on the x^2 term is understood to be 1. The coefficient on the x term (where here the exponent of one is understood) is -2, and the coefficient on the x^0 term is 1. The letters a, b, and c are often used to represent the x^2, x^1, and x^0 coefficients, so any general quadratic equation can be written as

$$ax^2 + bx + c = 0$$

Understand that a, b, and c can be positive, negative, zero, and are not necessarily integers. (However, it's sometimes easier to use the quadratic formula when the equation has integer coefficients, so it is common to rewrite equations so that they have all-integer coefficients.)

Example 7.3. Rewrite the following expression as a proper quadratic equation with integer coefficients, and identify a, b, and c.

$$x^2 - 4x = -3 - 2x + \frac{2}{5}x^2$$

Solution. If we want all integers for coefficients, we must algebraically eliminate the fraction in the last term. To do this, we multiply both sides of the equation by 5:

$$5\left(x^2 - 4x\right) = 5\left(-3 - 2x + \frac{2}{5}x^2\right)$$

$$5x^2 - 20x = -15 - 10x + 2x^2$$

In order to bring all terms to the left side of the equation, we subtract $2x^2$ from both sides, add 10x to both sides, and add 15 to both sides. We get

$$(5x^2 - 2x^2) + (-20x + 10x) + 15 = (-15 + 15) + (-10x + 10x) + (2x^2 - 2x^2)$$

$$3x^2 - 10x + 15 = 0$$

This is our proper quadratic formula with integer coefficients. This means that a = 3, b = -10, and c = 15. You should verify that this is correct.

Because any quadratic equation is an equation, it is expected that there is some number which, when substituted for x, makes the equation, in fact, equal zero. Consider the equation

$$x^2 - 2x + 1 = 0$$

If we substitute 1 for x, we have

$$1^2 - 2(1) + 1 = 1 - 2 + 1 = 0$$

That is, when 1 is substituted for x on the left side, evaluating each term and combining them **does** give zero as the final answer. The expression "x = 1" is called a <u>solution</u> to the quadratic equation. In chemistry, when we get quadratic equations in setting up some problems, our ultimate task is to find the solutions to the equation.

Solutions to quadratic equations can be almost any number. They can be positive or negative, integer or non-integer. They might even be <u>imaginary</u>, which is a number that involves the square root of -1. (The symbol i is used to symbolize $\sqrt{-1}$.) Imaginary numbers arise when you try to take the square root of negative numbers. For example, for the quadratic equation

$$x^2 + 1 = 0$$

where a = 1, b = 0, and c = 1, we solve by bringing the 1 to the other side of the equation:

$$x^2 = -1$$

Finally, we simply take the square root of both sides:

$$x = \sqrt{-1} \equiv i$$

The imaginary number i does not have physical significance, but it does have important mathematical significance.

In chemistry, however, we are dealing with things that do have physical significance, like amounts and pressures and concentrations of chemicals. Therefore, in chemistry, we limit allowable solutions to our quadratic equations as those solutions **that give physically measurable values.** This means that all of our final pressures and concentrations and amounts are positive (although depending on how we define our unknown x, its value may be negative), and **imaginary solutions are not allowed.** These restrictions simplify our attempts to find solutions using quadratic equations.

But how do we find solutions to quadratic equations? Not all of them will be solvable by just looking at the equation and plugging in simple numbers like 1 or 2 or -1. However, we have a useful tool: the quadratic formula. Any equation that has exponents on its variable has a maximum number of possible different solutions equal to the highest-numbered exponent. For quadratic equations, that means that there will be a maximum of two different solutions. (The word "different" is important, because both solutions may be the same. For the equation $x^2 - 2x + 1 = 0$, both of the solutions are x = 1.) For any general quadratic formula given by the general equation

$$ax^2 + bx + c = 0$$

the two roots x_1 and x_2 are given by the expressions

$$x_1 = \frac{-b + \sqrt{b^2 - 4ac}}{2a}$$

$$x_2 = \frac{-b - \sqrt{b^2 - 4ac}}{2a}$$

The "$\sqrt{}$" sign means "take the square root of the expression inside." In this case, the expression is "$b^2 - 4ac$." You should have a special key on your calculator for evaluating the square root of a number. (Roots are considered in a later section.) Keep in mind that a, b, and c might be negative numbers! Since the only difference in the two expressions is the + or - sign in front of the square root term in the numerator, the solutions are generalized using the ± symbol and are usually seen as

$$x = \frac{-b \pm \sqrt{b^2 - 4ac}}{2a}$$

Therefore, in order to determine the solutions to your quadratic equation, all you need to know are the coefficients of the equation, which has been written to equal zero.

Example 7.4. Show that $x = 1$ is the only unique solution to the quadratic equation $x^2 - 2x + 1 = 0$.

Solution. Since both solutions have $\sqrt{b^2 - 4ac}$ in them, we will evaluate that part first. Since $a = 1$, $b = -2$, and $c = 1$, we have

$$\sqrt{b^2 - 4ac} = \sqrt{(-2)^2 - 4 \cdot 1 \cdot 1}$$

$$= \sqrt{4 - 4} = \sqrt{0} = 0$$

We therefore have

$$x_1 = \frac{-(-2) + 0}{2 \cdot 1}$$

$$x_2 = \frac{-(-2) - 0}{2 \cdot 1}$$

Solving:

$$x_1 = \frac{2}{2} = 1 \quad \text{and} \quad x_2 = \frac{2}{2} = 1$$

The only unique solution is simply $x = 1$, which agrees with the solution we presented earlier.

The numbers won't always be as simple as in Example 7.4. However, the procedure for determining the solutions for a quadratic equation are the same **no matter how complicated the numbers look**. Therefore, once you have mastered using the quadratic formula, you can solve any quadratic equation. In chemistry, you should never get an imaginary number for your answer. They will always be real numbers.

Example 7.5. In working a problem involving a chemical reaction at equilibrium, the following equation is determined:

$$\frac{x^2}{0.02 - x} = 1.8 \times 10^{-5}$$

What are the possible values for x?

Solution. First, we should rewrite this equation into the proper format for a quadratic equation. We multiply through by 0.02 - x, and then bring all terms over to one side to get an equation that equals zero.

$$x^2 = (0.02 - x)(1.8 \times 10^{-5})$$

$$x^2 = 3.6 \times 10^{-7} - (1.8 \times 10^{-5})x$$

$$x^2 + (1.8 \times 10^{-5})x - 3.6 \times 10^{-7} = 0$$

This shows us that a = 1, b = 1.8×10^{-5}, and c = -3.6×10^{-7}. (Notice that c is negative!) We now plug these numbers into the quadratic formula:

$$x = \frac{-1.8 \times 10^{-5} \pm \sqrt{(1.8 \times 10^{-5})^2 - 4(1)(-3.6 \times 10^{-7})}}{2 \cdot 1}$$

$$x = \frac{-1.8 \times 10^{-5} \pm \sqrt{(3.24 \times 10^{-10} + 1.44 \times 10^{-6})}}{2}$$

$$x = \frac{-1.8 \times 10^{-5} \pm 1.2 \times 10^{-3}}{2}$$

By using the + sign in the numerator, we get

$$x = \frac{1.182 \times 10^{-3}}{2} = 5.9 \times 10^{-4}$$

By using the - sign in the numerator, we get the second solution:

$$x = \frac{-1.218 \times 10^{-3}}{2} = -6.1 \times 10^{-4}$$

Therefore, our solutions are x = 5.9 × 10⁻⁴ and x = -6.1 × 10⁻⁴. If these x's are representing solution concentrations, we need to recognize that it is not possible to have a **negative** concentration, and so we should omit the second solution. However, considerations like that will depend on the exact nature of the problem you are doing. Both answers are limited to two significant figures, and if you plug them into the original quadratic equation, you might find evidence of truncation error (i.e. the answer may differ from zero by a small amount).

Consult your textbook for specific chemistry problems that use the quadratic formula to solve for an unknown. You can find many of them in your text where chemical equilibrium is discussed.

Neglecting Terms

Earlier, when we set up the expression

$$\frac{x^2}{0.02 - x} = 1.8 \times 10^{-5}$$

we solved for x exactly by multiplying through with the denominator and expressing the equation as a proper quadratic equation. Consider the denominator, however: $0.02 - x$. We are subtracting some unknown amount, x, from 0.02. If the true value of the unknown x is very small compared to 0.02, then we are not affecting the absolute value of 0.02 all that much; in fact, if x is very small compared to 0.02, we are not going to change the final answer much if we simply neglect the x in the denominator:

$$\frac{x^2}{0.02 - x} = 1.8 \times 10^{-5}$$

neglect x **if it is small compared to 0.02**

In doing this, the algebra becomes a little simpler, and the final answer does not change much from the earlier answer:

$$\frac{x^2}{0.02} = 1.8 \times 10^{-5}$$

$$x^2 = (0.02)(1.8 \times 10^{-5})$$

$$x^2 = 3.6 \times 10^{-7}$$

$$x = 6.0 \times 10^{-4}$$

This answer is very close to the answer we got when we used the quadratic formula, and it was much simpler to determine.

So: when can we neglect a term like this? There are two rules:

- **Terms can only be neglected when they are being added or subtracted to another term.** You cannot neglect a term that is multiplying other terms.
- **Terms can only be neglected if they are small with respect to the other terms they are being added to or subtracted from.** How do you know if such terms will be small with respect to the other terms? Technically, you don't. You should always check your final answer for x and compare it to the term that is supposedly much larger than x. A good rule of thumb is that if x is expected to be much less than 10% of the larger value, you can save time by neglecting it and still come up with a reasonable final answer.

Of course, neglecting terms is an approximation that will ultimately change your final answer. But if the terms are quite different in magnitude, you can save yourself a lot of time by simplifying the mathematics. The following two examples show cases where neglecting the x works, and then doesn't work.

Example 7.6. Solve the expression for x by (a) using the quadratic formula, and (b) neglecting x and solving. Compare your answers.

$$\frac{x^2}{1-x} = 1 \times 10^{-10}$$

Solution. (a) We first rearrange the equation into a proper quadratic equation:

$$x^2 + 1 \times 10^{-10}x - 1 \times 10^{-10} = 0$$

Plugging into the quadratic formula, we find

$$x = \frac{-1 \times 10^{-10} \pm \sqrt{(1 \times 10^{-10})^2 - 4(1)(-1 \times 10^{-10})}}{2 \cdot 1}$$

Solving, we get

$$x = -1 \times 10^{-5} \text{ and } x = 1 \times 10^{-5}$$

(Technically, the answers are x = –0.000 010 000 1 and +0.000 009 999 9, but we are keeping our answers to one significant figure.)

(b) By neglecting x in the denominator, we get

$$\frac{x^2}{1} = 1 \times 10^{-10}$$

This is easily solvable as

$$x = 1 \times 10^{-5} \text{ and } x = -1 \times 10^{-5}$$

As you can see, neglecting the x in the denominator did not practically change our final answer. Keep in mind that you should check your final answer and compare it with the numbers in the problem to see if neglecting the x was justified!

Example 7.7. Solve the expression for x by (a) using the quadratic formula, and (b) neglecting x and solving. Compare your answers.

$$\frac{x^2}{1.00 - x} = 1.00 \times 10^{-1}$$

Solution. (a) By writing 1.00×10^{-1} as 0.100, we can save ourselves some work. Rewriting as a proper quadratic equation:

$$x^2 + 0.100x - 0.100 = 0$$

Using the quadratic equation:

$$x = \frac{-0.100 \pm \sqrt{(0.100)^2 - 4(1)(-0.100)}}{2 \cdot 1}$$

Solving, we find that

$$x = 0.270 \text{ and } x = -0.370$$

(b) Neglecting the x in the denominator, we get

$$\frac{x^2}{1.00} = 1.00 \times 10^{-1}$$

Again, this is relatively easy to solve; we simply take the square root of 0.100:

$$x = 0.316 \text{ and } x = -0.316$$

We are off by about 15% on one solution and 16% on the other. These answers would probably not even be considered correct on a quiz or exam, because they are so far off (especially when compared to the previous example!). In this case, our unknown x is not less than 10% of the number it is subtracted from in the original problem, and neglecting the x was not the appropriate thing to do.

Students always have the question, "Should I neglect the x?" Unfortunately, this is one question where a simple answer won't work, in spite of the 10 Percent Rule mentioned above. Theoretically, it is best to try to solve the complete quadratic equation. Practically, in time and with practice, you will develop the sophistication to know when to neglect a term and when to not. Ask your instructor or consult your textbook for specific chemistry problems that require such decisions.

Roots

Most people are familiar with square roots (in fact, we have already used them in problems). The concept of a square root is, "What number 'x' times itself equals another number 'y'?" The number x is said to be the square root of y. It is written as

$$x = \sqrt{y}$$

For example, the square root of 9 is 3:

$$3 = \sqrt{9}$$

We know this because we can square both sides of the equation and get an equality again:

$$3^2 = 9$$

Negative numbers can also be square roots. (However, the square root of negative numbers brings up imaginary numbers, which we will not consider here.) Since we understand that $(-3)^2$ also equals nine, then it is proper to think of -3 as the square root of 9 also:

$$-3 = \sqrt{9}$$

Technically, when taking a square root of a number, **the positive and negative square root must be considered.** In many cases involving physical quantities, the negative square root is not an acceptable answer, but you should keep in mind that mathematically square roots can be positive **and** negative. Many people already know the numbers that have integral square roots (i.e. 1, 4, 9, 16, 25, 36, etc.), but all numbers have square roots. They are just not integers, and usually are evaluated using a calculator. For example, the square root of 10 is 3.162 277 6......

We can also represent square roots using exponents. Since we designate the square of a number with a superscripted "2", we designate the square root of a number using the reciprocal of 2: $\frac{1}{2}$. Thus, $\frac{1}{2}$ is used as a superscript, or an exponent, to indicate a square root. For example,

$$10^{1/2} = \sqrt{10} = 3.162\ 277\ 6.....$$

The rules of positive and negative exponents hold for these fractional exponents, too.

Example 7.8. What is the value of $10^{-1/2}$?

Solution. From the rules of negative exponents, $10^{-1/2}$ equals

$$\frac{1}{10^{1/2}} = \frac{1}{\sqrt{10}} = \frac{1}{3.162\,277...} = 0.316\,277...$$

Can you evaluate this on your calculator properly?

A square root is based on a squared number, a number multiplied by itself. A square root is the inverse operation of a square, so that taking a number, squaring it, and then taking the square root of the resulting number regenerates the original number (and vice versa). So,

$$x = \left(\sqrt{x}\right)^2 = \sqrt{x^2} = x$$

A square and a square root cancel each other.

The same thing holds true for other, larger powers. They, too, have their inverse. For example, the inverse of a number raised to the third power, or cubed, is called a <u>cube root</u> and is indicated by the symbol $\sqrt[3]{}$. So, for example, if 3 x 3 x 3 = 27, then the cube root of 27, or $\sqrt[3]{27}$, is 3. Again, most numbers have non-integer cube roots, and we usually evaluate them using a calculator. Other roots – fourth roots, fifth roots, etc. – are also possible but are rare in chemistry.

Cube roots can also be written as fractional exponents. Since we write a cube with a 3 as a superscript, we indicate a cube root as a $\frac{1}{3}$ as a superscript. For example, $x^{1/3}$ implies the cube root of x. Cube roots cancel an exponent of 3, just like square roots cancel an exponent of 2.

Many calculators have a square key as well as a square root key: $\mathbf{x^2}$ and $\sqrt{}$, usually. However, most calculator don't have specific keys for cube roots or larger roots. Some calculators have keys that look like $\sqrt[\square]{\square}$, where you have to enter the root (the box outside the sign) and the number you're taking the root of (under the sign). On other calculators, you have to use a key that looks like $\mathbf{y^x}$ and enter the original number and the root in fractional form (i.e. **0.5** for a square root, **0.333 333 333 3...** for a cube root, **0.25** for a fourth root, etc.). Know in advance exactly how to do cube and other roots on your calculator! Exercises at the end of this chapter will require you to practice these skills.

Example 7.9. Solve for x from the following expression.

$$\frac{0.351}{x^3} = 9.44 \times 10^{-6}$$

Solution. Rearranging the expression to isolate x^3 on one side of the equation in the numerator:

$$\frac{0.351}{9.44 \times 10^{-6}} = x^3$$

Evaluating the fraction, we find that

$$x^3 = 3.72 \times 10^4$$

Taking the cube root, we find that

$$x = 33.4$$

You should check this with your own calculator to see if you are doing cube roots correctly. If you do not get that answer, consult your calculator manual or your instructor.

You will occasionally find cube roots in chemistry problems, but rarely roots higher than that. However, hopefully this introduction to roots will allow you to apply this knowledge should you come across them.

Exponentials and Logarithms

In considering the function "x^2," the "x" part is called the <u>base</u> and the "2" part is called the <u>exponent</u>. Such functions, generally called <u>power functions</u>, have the variable as the base and a number as the exponent.

Suppose you have it the other way around. Suppose a number is the base and an expression of some variable(s) is the exponent. The expression "2^x" would be an example. This expression is called an <u>exponential function</u>, or simply an <u>exponential</u>. (Notice the similarities in the names "exponent" and "exponential.")

The two common bases in chemistry are 10 and a number symbolized by e. The number e is like π; it has a particular value of an infinite number of digits. The numerical representation of e is 2.718 281 828 46....., so you see why it is just easier to use the letter e to represent it.

We have already met some applications of exponentials. First, when we discussed scientific notation earlier, we were using the so-called "base 10" system of exponentials:

$$10^0 = 1$$
$$10^1 = 10$$
$$10^2 = 100$$
$$10^3 = 1000$$
etc.

We also found that we could take square roots, cube roots, and other roots by expressing them as fractional exponentials of numbers. As such we can calculate the square, cube, and fourth root using the **10ˣ** key on our calculator:

$$10^{1/2} = 3.162\ 227\ ....$$
$$10^{1/3} = 2.154\ 434\ ....$$
$$10^{1/4} = 1.778\ 279\ ....$$

and so forth. In most cases, you enter the exponent into your calculator first, then press the **10ˣ** key. You should verify the above three roots to see if that is the case for your calculator.

The "base e" exponential , also called the <u>natural exponential</u>, is used the same way. There is usually a key that looks like **eˣ** on most calculators. After entering the correct number for the exponent, pressing the **eˣ** key will raise 2.718 281 828 46.... to that power.†

Example 7.10. Use your calculator to evaluate the expression

$$10^{(6\cdot3)/(2\cdot11)}$$

Solution. Remembering the proper order of operations, you should get 6.579 332 .... on your calculator. Did you?

Just as powers and roots were mathematical inverses, exponentials have inverse operations as well. They are called <u>logarithms</u>. Since we are focusing on 10 and e as bases for our exponentials, we will concentrate on the logarithms that are the specific opposite of these exponentials. They are abbreviated log and ln. (The second logarithm is understandably called the <u>natural logarithm</u>.)

---

† The numerical value for e may seem strange, but there are definite mathematical reasons for it. Most people don't realize how many times e appears in nature (and hence its name, "natural exponential"). The curvature of the nautilus shell, the decrease in the pressure of the atmosphere with altitude, the variation of molecular speeds of gas molecules with temperature – all are related to e. Neat, huh?

It is crucial to not confuse the two logarithms. You will not get the same numerical answer when you perform a base-10 logarithm as when you perform a natural logarithm. Keep in mind that when we use the word "logarithm," we mean the **base-10** logarithm. If we use the *e*-related logarithm, we always state it explicitly as "**natural** logarithm."

How are logarithms related to exponentials? The logarithm of a number is that power to which the base is raised to generate that number. Pictorially,

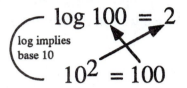

We speak of it as "the logarithm of 100 is 2, because 10 raised to the power of 2 equals 100." Needless to say, the log 10 = 1, because $10^1 = 10$.

All positive, non-zero numbers have logarithms. (Zero and negative numbers do not.) As with roots, the logarithms of most numbers are not integers. For example, the natural logarithm of 10 is

$$\ln (10) = 2.302\ 585 \ldots$$

Remember what this means: $e^{2.302\ 585\ \ldots} = 10$. Can you show that this is true on your calculator?

Most of the time we meet exponentials and logarithms in chemistry, they are part of equations or formulas. For example, in chapter 6 we saw an equation

$$E = E° - \frac{RT}{n\mathcal{F}}\ln Q$$

where E is a voltage of a battery, E° is the battery's standard voltage, R and T have their usual definitions, n is the number of moles of electrons transferred in the balanced chemical reaction, $\mathcal{F}$ is the Faraday constant, and Q is what is called the reaction quotient, which is the typical "products-minus-reactants" expression. In this equation, you have to evaluate the natural logarithm of the number given by the reaction quotient Q. That natural logarithm, which is a pure number with no units, is part of the equation.

One point about logarithms and exponentials: the number you take the logarithm of, or the exponent in the exponential, are **unitless**. You can only take the logarithm of a pure number. It makes no sense to evaluate "log (10 km)," since the logarithm of a unit does not exist. Similarly, when evaluating exponentials, the expression that is the exponent must be unitless overall. The next example illustrates this idea.

Example 7.11. Some exercises require that you evaluate the following expression:

$$e^{-E_A/RT}$$

where E_A is an energy, R is the ideal gas law constant, and T is the absolute temperature. Evaluate the exponential for E_A = 25.09 kJ/mol, R = 8.314 J/mol·K, and T = 298 K.

Solution. By plugging in the values, we see the expression becomes

$$e^{\frac{-25.09 \text{ kJ/mol}}{(8.314 \text{ J/mol·K})(298 \text{ K})}}$$

The units "mol" and "K" will cancel, but what about the energy units? They are different. One of them must be converted to another unit in order for these last units to cancel. Let us convert the kJ units into J:

$$25.09 \text{ kJ} = 25{,}090 \text{ J}$$

We substitute and get

$$e^{\frac{-25{,}090 \text{ J}}{(8.314 \text{ J})(298)}}$$

and now the "J" units in the expression for the exponent will cancel. We can now evaluate the exponential:

$$e^{\frac{-25{,}090}{(8.314)(298)}} = e^{-10.1268...} = 4.00 \times 10^{-5}$$

The final answer is a pure number; there are no units. Plug the above numbers into your calculator and see if you get the same final answer.

Remember that logarithms and exponentials are inverses: one cancels the effect of the other. Using x as our variable, this means that

$$\log (10^x) = 10^{\log x} = x$$

Since this is so, suppose you have to find the value of an unknown that's inside a logarithm or part of an exponent of an exponential? You will have to take the exponential of both sides of the equation, or take the logarithm of both sides of the equation, in order to isolate your unknown. For example, suppose you have the following equation

$$\Delta G = -RT \ln K$$

and you are given that $\Delta G = 1.86$ kJ/mol at a temperature of 298 K and you need to find the value of the variable K. (Do not confuse it with the symbol for degrees Kelvin. Remember: know what your variables stand for!) R, the ideal gas law constant, is 8.314 J/mol·K. We plug in:

$$1.86 \text{ kJ/mol} = -(8.314 \text{ J/mol·K})(298 \text{ K}) \ln K$$

The Kelvin and mole units cancel, but the energy units are again inconsistent. Again, we will convert kJ to J and have $\Delta G = 1860$ J/mol:

$$1860 \text{ J} = -(8.314 \text{ J})(298) \ln K$$

Now we can cancel the energy units from both sides of the equation. Notice that, at this point, there are no units left.

Remember the tactic for solving for an unknown variable: it must be by itself on one side of an equation and in the numerator. If we are looking for K, we bring all of the numbers to one side of the equation to get

$$0.751 = \ln K$$

The natural logarithm is an **operation**; we cannot simply just divide by "ln" and isolate K all by itself. This equation is asking "The natural logarithm of what number K is equal to 0.751?" In order to figure this out, we must perform the **inverse operation** in order to get rid of the logarithm. Well, the inverse operation is the exponential, and in this case it is the natural exponential e. Therefore, we raise e to both sides of the equation:

$$e^{0.751} = e^{\ln K}$$

Since exponentials and logarithms are inverses, they cancel each other out, and the right side of the equation simply equals K. Therefore,

$$K = e^{0.751}$$

We can evaluate this using the e^x key on a calculator. We enter **0.751**, press the e^x key, and get our final answer:

$$K = 2.119$$

Notice how we isolated K all by itself by first isolating ln K and then performing the inverse operation on both sides of the equation. Why didn't we use the 10^x key? Because the "ln" means that we are dealing with the **natural** logarithm, not the base-10 logarithm. If you use the wrong exponential to inverse a logarithm, you will get the wrong numerical answer.

What are the rules regarding significant figures and logarithms? The most common time to consider significant figures is when you are taking a log (i.e. a base-10 logarithm) of a small number in calculating pH of aqueous solutions. The pH is defined as

$$pH = -\log [H^+]$$

where $[H^+]$ is the hydrogen ion concentration in units of molarity (which is the understood unit; it is not included inside the logarithm).

When you consider the relationship between an exponential number and its resulting logarithm, the numbers to the **left** of the decimal point of the resulting logarithm are related to the exponent of the exponential, whereas the numbers to the **right** of the decimal point of the resulting logarithm are related to the mantissa of the exponent (see Chapter 1 for the definitions of mantissa and exponent). Significant figures for logarithms are therefore limited by the significant figures of the **mantissa. A proper logarithm has as many decimal places as the mantissa has significant figures.** There is no restriction on the number of significant figures <u>before</u> the decimal point.

For example, suppose we consider the significant figures after evaluating

$$- \log (3.66 \times 10^{-6})$$

Remember that the "$\times 10^{-6}$" only places the 3, 6, and 6 in the proper decimal positions. The logarithm, negated, is

$$5.436\ 518\ 914\ ....$$

(Is this what you get when you take the logarithm of the original number and then negate it? If not, you may be working your calculator improperly.) The number to the left of the decimal point, 5, is indicative of the exponent on the exponential and does not need to be considered. The original number had a mantissa with three significant figures. Therefore, we limit our decimal figures to three places. In this case, we round up to get for our final answer

$$- \log (3.66 \times 10^{-6}) = 5.437$$

The relationship between significant figures and decimal places is also valid when going in the opposite direction, too. For example, 10 raised to the 5.224 power is

$$10^{5.224} = 1.67 \times 10^5$$

In evaluating this, we limit our mantissa to three significant figures, because our power has three decimal places. Because of the properties of logarithms, if we took the logarithm of 1.67×10^5, we would not get 5.224 back, due to truncation errors. However, we hardly ever have to take multiple logarithms and exponentials in the same mathematical problem, so such errors do not usually affect the final answer much.

Student Exercises

Be careful when using your calculator! If you are not getting the correct answers, it may not be that you don't understand the math. Instead, you may not be operating your calculator correctly. If you have questions, consult your manual or see your instructor.

7.1. Write the following expressions as fractions having all positive exponents. If a variable appears more then once in an expression, be sure to combine the exponents properly.

(a) $x^5 + y^{-3}$

(b) $\dfrac{v^3 x^{-2}}{z^4 v^{-1}}$

(c) $1 - \dfrac{y^3}{\left(\dfrac{1}{y^{-3}}\right)}$

(d) $\dfrac{2^8}{2^6}$

(e) $\dfrac{10^2 \cdot 3^4}{10^5 \cdot 3^{-3}}$

(f) $10^5 \cdot 10^4 \cdot 10^{-8}$.

7.2. In your home, electricity use is measured in terms of a unit called "kilowatt-hours." A similar unit composed of more basic units is the "watt-second." From the definition of the watt, write this derived unit in terms of fundamental units written with positive exponents. What is this combination of units equal to?

7.3. The units on Planck's constant, a fundamental universal constant, are J·sec. Show that this is equal to the units for angular momentum, which defined as (mass)(velocity)(distance). HINT: What are the units used to describe mass, velocity, and distance?

7.4. Determine solutions to the following quadratic equations.

(a) $x^2 - 6x + 9 = 0$.

(b) $x^2 - 16 = 0$.

(c) $a^2 - 5a + 6 = 0$.

7.5. Determine solutions to the following quadratic equations.

(a) $3x^2 + 24x - 22 = 0$.

(b) $0.8x^2 - 4.2x - 2.7 = 0$.

(c) $-6x + 6x^2 = 12$

(d) $\dfrac{4x^2}{0.5 - x} = 10$

7.6. Find solutions to each of the following expressions by neglecting the appropriate terms and compare your answer to solutions found using the quadratic formula. Were you justified in neglecting the terms you did?

(a) $\dfrac{2x^2}{0.0400 - x} = 3.74 \times 10^{-2}$

(b) $\dfrac{x^2}{0.200 - x} = 8.3 \times 10^{-5}$

7.7. Write down the following expressions using fractional exponents and evaluate them using a calculator.

(a) The cube root of 10.

(b) The fourth root of 25.

(c) The cube root of 1000.

(d) The reciprocal of the fifth root of 250.

(e) The reciprocal of the square root of 2.

(f) The negative of the sixth root of 1.01.

7.8. Evaluate the following expressions using your calculator.

(a) $\sqrt[3]{6^{1.5}}$

(b) $\left(\sqrt{15}\right)^2$

(c) $\dfrac{\sqrt{16}}{\sqrt{4}}$

(d) $\dfrac{\sqrt[3]{10}}{\sqrt[4]{10}}$

7.9. Solve the following equations for the unknown variable.

(a) $\dfrac{56.0}{x^3} = 1.44 \times 10^{-2}$

(b) $\dfrac{4977}{x^3} = \dfrac{22\,844}{x^5}$

(c) $x = e^{-569/444}$

(d) $2.964\,23 = 10^x$

(e) $2.1116 \cdot \ln\left(\dfrac{x^2}{0.654}\right) = -0.4482$

(f) $e^{\ln 8.883} = x$

(g) $\log\left(\dfrac{10^3 \cdot 10^{-5}}{10^4 \cdot 10^2}\right)$

(h) $1.077 = 1.100 - \dfrac{(8.314)(295)}{(2.00)(96,500)} \ln Q$

7.10. Most people don't memorize the value for e, since it is easy to get from your calculator. Can you get your calculator to display the value for e? There are several ways, depending on the model of calculator you have.

7.11. What is the pH of a solution that has $[H^+] = 4.208 \times 10^{-3}$? Use the equation $pH = -\log[H^+]$, and express your answer to the proper number of significant figures.

7.12. What is the $[H^+]$ of a solution that has a pH of 10.882? Express your answer to the correct number of significant figures.

7.13. Throughout the course of this book, we have discussed several mathematical operations that are "opposite" or "the inverse" of each other. Name four pairs of operations that can be considered "opposites."

Answers to Student Exercises

7.1. (a) x^5y^3 (b) $\dfrac{v^4}{z^4x^2}$ (c) 0 (d) 2^2, or 4 (e) $\dfrac{3^7}{10^3}$, which equals 2.187 (f) 10^1, or 10.

7.2. A watt-second equals $\dfrac{kg \cdot m^2}{sec^2}$, which is equal to a joule. Watt-seconds are therefore a measure of energy.

7.3. Both sets of units can be shown to equal $\dfrac{kg \cdot m^2}{sec}$.

7.4. (a) $x = 3$ (b) $x = 4$ and $x = -4$ (c) $a = 2$ and $a = 3$.

7.5. (a) $x = 0.830\ 458\ ...$ and $x = -8.830\ 458\ ...$ (b) $x = 5.829\ ...$ and $x = -0.579\ ...$ (c) $x = 2$ and $x = -1$. (d) $x = 0.427\ ...$ and $x = -2.927\ ...$

7.6. (a) Neglecting the x in the denominator of the original expression, we find $x = \pm\ 0.027\ 349\ ...$, while using the quadratic formula, we find $x = 0.019\ 553\ ...$ and $x = -0.038\ 253\ ...$ Neglecting the x does not give a good approximate answer. (b) Neglecting the x: $x = \pm\ 4.074\ 309\ ... \times 10^{-3}$, while using the quadratic formula, $x = 4.033\ 021\ ... \times 10^{-3}$ and $x = -4.116\ 021\ ... \times 10^{-3}$. This approximation was justified.

7.7. (a) $10^{\frac{1}{3}} = 2.154\ 434\ ...$ (b) $25^{\frac{1}{4}} = 2.236\ 067\ ...$ (c) $1000^{\frac{1}{3}} = 10$ (d) $\dfrac{1}{250^{\frac{1}{5}}} = 250^{-\frac{1}{5}} = 0.331\ 445\ ...$

(e) $\dfrac{1}{2^{\frac{1}{2}}} = 2^{-\frac{1}{2}} = 0.707\ 106\ ...$ (f) $-\left(1.01^{\frac{1}{6}}\right) = -1.001\ 659\ 76$.

7.8. (a) $2.449\ 489\ 742\ ...$ (b) 15 (c) 2 (d) $1.211\ 527\ ....$

7.9. (a) 15.7 (to three sig figs) (b) $\pm\ 2.142$ (four sig figs) (c) 0.278 (d) 0.471 91 (e) 0.727 (273 973 ...) (f) 8.883 (g) -8 (h) $Q = 6.11$

7.10. To get the value for e, enter **1** into your calculator and hit the $\mathbf{e^x}$ key. If your calculator does not have an $\mathbf{e^x}$ key, you may have to hit the inverse natural logarithm, probably **INV**, then **ln**. Which way works for your calculator?

7.11. pH = 2.3759.

7.12. $[H^+] = 1.31 \times 10^{-11}$, where the unit is assumed to be molarity.

7.13. The following pairs of operations can be considered opposites or inverses of each other:

addition & subtraction

multiplication & division

powers and roots

logarithms and exponentials

Chapter 8. Making Graphs

Introduction

Graphs are commonly used in chemistry and other sciences to visually represent related sets of information. In a glance, they can illustrate information in a way that tables of data can't, and often they are the best way to illustrate trends and relationships between different observable parameters of a chemical system. Many times – especially in a laboratory course – you will have to take data that you yourself measured and plot them on a graph to see what trend or relationship exists. It is important that the graph be represented properly for the best possible understanding of the information. A bad graph, with data poorly plotted, is a waste of graph paper more than it is an effective tool for communicating information.

In this chapter, we will review some of the basic concepts for constructing good graphs. The way to set up a graph depends on what you want it to show, and there are a few choices you will have to make in order to present your data effectively or to determine additional information from your graph. Because of the wide range of possibilities for setting up a graph, we will focus on one major example and develop it, instead of presenting multiple examples throughout the chapter.. Hopefully, you will be able to apply the ideas for constructing proper graphs to the student exercises by the end of the chapter.

Straight or Curved Lines?

When you plot data on a graph, you are using data that are related to each other somehow, and the graph is a visual representation of that relationship. For example, if you have a fixed amount of gas at a certain temperature, the volume of that gas is related to its pressure. (This relationship is called Boyle's law.) If you made measurements of the pressure of gas at different volumes, you could plot the different values of pressure versus the volume and produce a graph.

In almost all cases of related data, the change in one variable with respect to the other is **smooth**. That is, there are few abrupt and sudden changes, unless something abrupt is done to your system (like dropping a hot piece of metal into cool water and measuring the resulting temperature of the water). What this means is that most of your graphs will look like the following two:

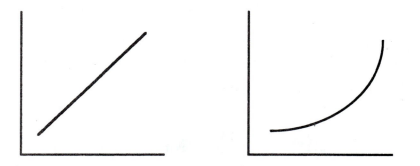

You will not see graphs that have sudden, abrupt changes like this:

This graph is not <u>smooth</u>.

Given that most of our plots will represent smooth changes in measurements, one of the keys to understanding a graph is to know ahead of time whether your graph is supposed to represent a straight line or a curved line. Straight lines are easy to deal with because mathematics has a simple equation that can be used to understand any straight line. This equation is

$$y = mx + b$$

where x and y are the two variables that are related to each other, m is the <u>slope</u> of the line, and b is called the <u>y-intercept</u>. The y-intercept b is the point where the straight line crosses the y axis (the vertical axis), which is the point on the x axis where x = 0. The slope m represents how steep the straight line is. A flat, horizontal line has a slope of zero, while a perfectly vertical line has an infinite slope. Slopes can be positive or negative. A positive slope means that the line goes up in the y dimension as it goes from left to right in the x dimension. A line with a negative slope goes down in the y dimension as it goes from left to right in the x dimension. The y-intercept can also be positive or negative.

Because straight lines are easy to understand, many equations in chemistry are written in such a form that the two variables, when plotted, give a straight line. One example is the relationship between a rate constant k, which tells how fast a chemical reaction proceeds, and the temperature of the chemical reaction T. There is a simple exponential equation that relates the two:

$$k = A \cdot e^{-E_A/RT}$$

In this equation, k is the rate constant, A is a proportionality constant, E_A is called the <u>activation</u> <u>energy</u> of the reaction, and R and T have their usual definitions. This equation relating k and T will not give a straight line if k and T are plotted on a graph. However, if we take the natural logarithm of both sides, we get

$$\ln k = \ln\left(Ae^{-E_A/RT} \right)$$

$$\ln k = \ln A + \ln e^{-E_A/RT}$$

$$\ln k = \ln A - \frac{E_A}{RT}$$

In the second step, we are using the property of logarithms that says that $\ln(xy) = \ln x + \ln y$, and in the final step we are using the fact that the natural logarithm and the natural exponential are inverses, so they cancel each other (see the previous chapter). This last equation is written in the form of a straight line if we consider one of our variables as $\ln k$ and the other variable as $\frac{1}{T}$:

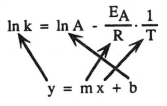

Therefore, if we plot $\ln k$ on the y axis and $\frac{1}{T}$ on the x axis, we should get a straight line with a slope equal to $-\frac{E_A}{R}$ (notice that it is negative) and a y-intercept of $\ln A$. (Notice that the terms are not in the same order as "y = mx + b" but are still identifiable as slope and intercept.) Many equations in chemistry can be manipulated mathematically, like we did above, to change the equation into a form that will yield a straight line when the variables are plotted properly.

Two related variables can also be plotted and give a curved line. One example is the relationship between pressure and volume of a fixed amount of gas at a given temperature. The mathematical equation that relates pressure P and volume V is

$$P = K \cdot \frac{1}{V}$$

where K is a proportionality constant. This equation suggests that a plot of P versus $\frac{1}{V}$ would give a straight line, and it does. But it is more common to plot P versus V directly, and when you do, you get a graph that looks like

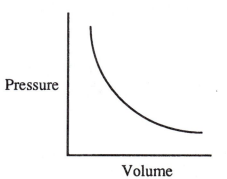

In this case, the graph is a smooth curve, not a straight line. (Notice that we have labeled our axes for the first time. We will be discussing this shortly.)

In short, it's a good idea to know ahead of time whether your graph is expected to be a straight line or a curved line. No graph, however, is ever going to be a **perfect** straight or curved line. That's because when you are working with experimental data, there will always be experimental error. This experimental error usually adds a little bit of imperfection, or <u>scatter</u>, to your graph. But the trend in the data plotted in a graph should still be obvious: either a straight line or a curved line.

Being able to recognize that an equation should yield a straight line is an important skill. The following example gives some idea of how to spot equations that can be graphed to give a straight line.

Example 8.1. Each of the following equations will yield a straight line, if plotted correctly. For each given set of variables, indicate what to plot as y, what to plot as x, and identify the correct expression for m and b. Assume that all other variables are constants. (a) ln A – ln K = –kt, where the variables are A and t. (b) $\dfrac{1}{A} = \dfrac{1}{K} + kt$, where the variables are A and t. (c) pH = pK + log(ratio), where pH and "ratio" are the variables.

Solutions. (a) If we rewrite the equation to

$$\ln A = -kt + \ln K$$

then we would plot ln A as y and t as x. The slope would be -k, while the y-intercept would be ln K. (b) For this equation, $\dfrac{1}{A}$ would represent y, t would be our x, the slope would be k, and the y-intercept would be $\dfrac{1}{K}$. (c) The pH would be the y variable, and the "log(ratio)" would act as the x variable. There is no visible constant multiplying the "log(ratio)" term, but there is always an understood 1 as a multiplier. Therefore, the slope of this equation is equal to 1. The y-intercept will be equal to pK.

Axes

When graphing the behavior of one variable versus another, you commonly start out with a set of data, usually in the form of a table. Usually, one of the variables listed in the table has values that were determined by the experimenter, and the other variable is the property that was measured. The variable that is determined is called the independent variable, while the variable that is measured is the dependent variable. Graphs are plots of how a dependent variable changes with respect to the independent variable.

Most graphs are set up by using two perpendicular scales, one for each variable, called axes (singular axis). Some graphs use a circular grid to plot data, but all of the cases we will be considering will use two perpendicular axes. Typically, the horizontal axis is considered the x axis, and the independent variable is measured against this axis. The vertical axis is referred to as the y axis, and the dependent variable is measured against this axis:

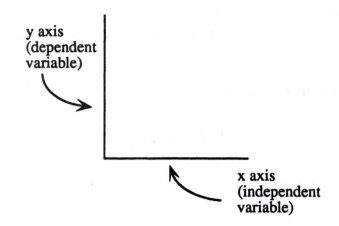

If you use graph paper you bought in a store, you will probably notice that each page isn't completely covered with a grid of lines (which helps in plotting your data). Some of the lines of the grid are usually darker than others. This helps you keep track of your position. We will start presenting graphs on a grid and will do so for the remainder of this chapter.

When you set up a graph, you should first draw the axes yourself with a pen or pencil. You can either draw them on the edge of the grid of lines, or you can indent your graph by drawing axes independent of the grid's edge. The overall look of your graph should not be affected by exactly where you draw your own axes – as long as you give yourself an appropriate amount of space for all of your data. (WARNING: Your instructor may require that you draw your axes in a particular way. You should check with your instructor regarding the placement of axes on a sheet of graph paper.) For example, you might have a set of axes that look like this:

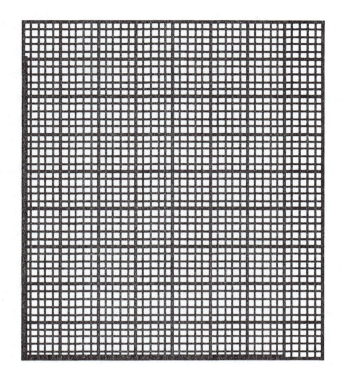

Or, you may choose to "indent" your axes so they aren't on the edge of the grid:

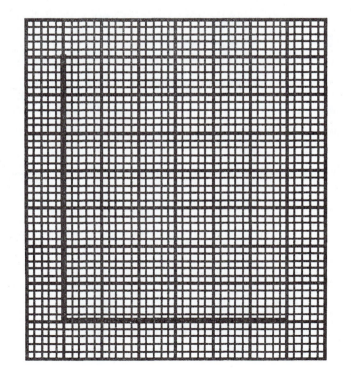

Again, you should check with your instructor to find out whether one way of drawing axes is preferred. Many graph papers have grids showing dark and light lines, like the examples above. Each dark line in these examples is five marks away from the next dark line, and four lighter lines separate them. If the axes you draw are indented, they are usually drawn over the darker grid lines on the graph paper. (Graph paper comes with grids marked out in English units, metric units, and even logarithmic units, but they all usually have dark and light grid lines to help keep track of your position along an axis.) Axes sometimes represent a zero value for one of the variables. If this is the case for your particular graph and some of your data are negative numbers, you may have to indent your axes in order to accommodate all of the data in your plot.

Numbering and Labeling Axes

For beginning students, it seems that the most difficult part of setting up a graph is how to number and label the axes, in order to give each axis a scale. This is because the scales for the axes of every graph depend on the data, and when different data are being plotted, the scale for that graph will be different. What this means is that there are no truly absolute rules for determining the scales for the axes. However, there are some general guidelines that will make any graph a more acceptable visual presentation of your data.

As mentioned, the grids on graph paper can be marked out in metric units or English units. In both cases, the intervals between the consecutive parallel grid lines are the same. These are linear scales, and each interval is meant to represent the same change in the variable's value. There is also a logarithmic scale, where each major interval (usually marked by the darker grid lines or slightly larger tick marks) represents a power of ten. This means that each successive major interval is ten times the previous one. This allows you to plot small and large numbers on the same axis, but the scale is completely different. The following shows an example of a linear scale versus a logarithmic scale for one axis:

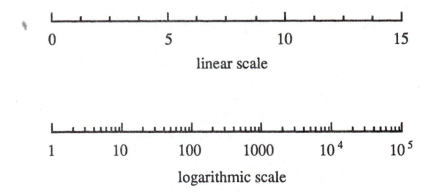

linear scale

logarithmic scale

If both axes are logarithmic, the graph is referred to as a "log-log graph." In chemistry we almost exclusively use linear plots.

Good graphs should use <u>most of the area bound by the axes</u>. Furthermore, the area bound by the axes should use most of the area of the page of graph paper. These two ideas allow you to make the most efficient use of space for the graph. The axes drawn on the previous page use almost all of the area of the grid on the graph paper. As a poor example, consider the following:

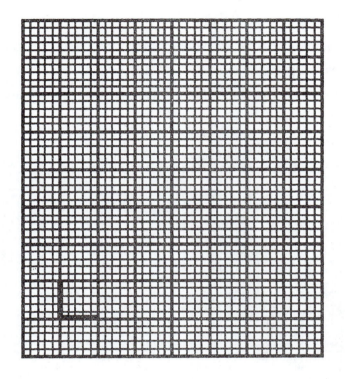

In this case, axes are drawn but only cover a fraction of the total area of the available graph paper grid. If you were to try to plot a collection of points in such a small area, any trend in your data would be hard to see.

Drawing the axes for a graph and determining what values each grid line should have cannot be done without consideration of the data. Most axes start and end on a dark grid line, if the graph paper has them. These darker lines are the lines of the grid that are usually marked to indicate specific values of the variable being plotted on that axis. The extreme values represented by the first and last interval <u>should include the entire range of values</u> for the dependent and independent variables. Each major interval (i.e. the distance between labeled lines) <u>should have the same change in value</u> (for linear graphs). This means that you have to consider your data, the size of your graph, and the number of major intervals each axis will have. Usually each interval has a simple change in value, like 1 or 3 or 5 or 10. Good graphs don't have intervals having odd values, like 5.77 or 0.054 22 or 10.6209.

As an example, consider the following data collected for a gas at constant volume:

Temperature, K	Pressure, atm
100	0.200
200	0.400
250	0.500
300	0.600
350	0.700
400	0.800

Suppose we want to plot the independent variable temperature versus the dependent variable pressure. Temperature is on the x axis and pressure is on the y axis. Temperature ranges from 100 K to 400 K, and the pressure ranges from 0.200 atm to 0.800 atm. Suppose, too, that we are working with a piece of graph paper like so:

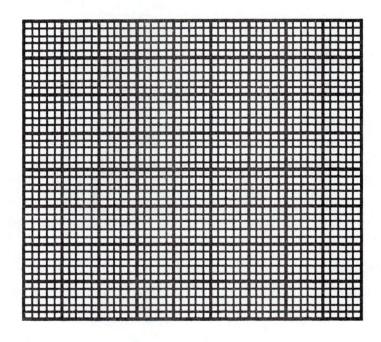

We have a total of nine major intervals for our x axis, which will be for temperature, and a total of 8 major intervals for our y axis, which will be for pressure. There are many possible ways of setting up this graph; what follows is not the only way. Consider the y axis first. If each major interval were equal to 0.1 atm, then we can use the edge of the grid and label each dark line from 0 to 0.8 atm in 0.1-atm increments. We will be able to include all of our pressure data on that scale. Our y axis for our graph will look like this:

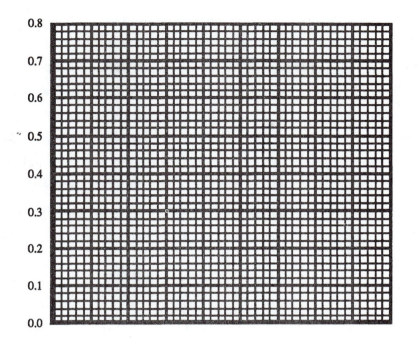

For the x axis, if we have each major interval represent 50 K, we can go from 0 K to 450 K. If we include our x axis, the graph would now look like this:

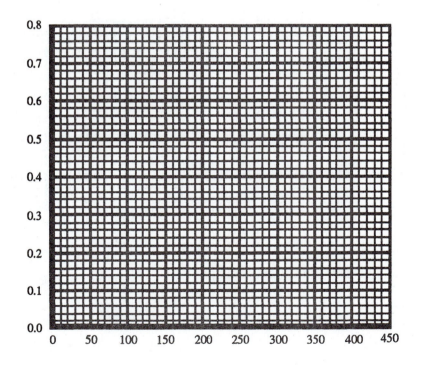

While it may seem that we are now ready to plot the data points and finish our graph, there is something missing. Since a graph is supposed to be a concise pictorial representation of data, the graph must include what the numbers stand for. This is crucial, because graphs are useful for illustrating principles to other people, and these other people won't necessarily know what is being plotted with respect to what! **Each axis of the graph must be labeled with the appropriate variable <u>name and unit</u>.** Both variable name and unit are necessary, especially since many variables can be expressed in different units. In this case, the x axis is the temperature in units of degrees Kelvin, and the y axis represents pressure in units of atmospheres. Many times, a graph is also given a title so that people can differentiate between graphs. The title is often written outside the grid on the page if there is room, or directly on the grid out of the way of any of the data points. The labels and title must be included in a proper graph, like so:

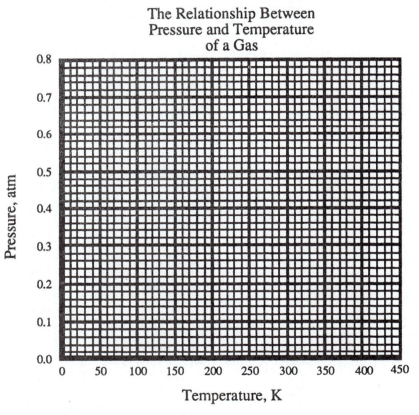

We are now ready to take our data and construct the plot.

Example 8.2. Shown below is a different grid that has been set up to plot the temperature/pressure data from the table above. Although the grid is somewhat smaller than the example we worked out above, there are some deficiencies in how the axes were set up. What do you think could be done better?

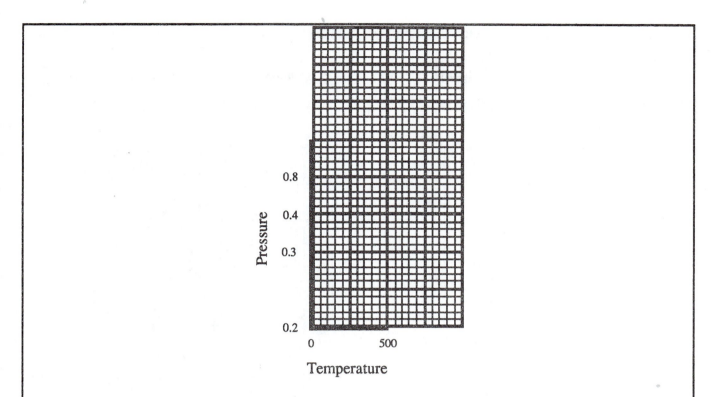

Temperature

Suggestions for Improvement: The x axis is drawn too short, and the temperature values won't be spread out enough for any relationship to be noticeable. There are no units on either axis. The y axis doesn't extend as far as it could, and the numerical labels for the major intervals aren't equally spaced. The y axis could be extended so that each interval is 0.1 and all of the pressure data could be plotted. The graph does not have a title. There are many possible ways to set up a better representation to graph the given data.

Plotting the Points

Now that the axes for the graph have been set up, the individual data sets can be plotted. Let us recall the data we are plotting:

Temperature, K	Pressure, atm
100	0.200
200	0.400
250	0.500
300	0.600
350	0.700
400	0.800

You need to recognize which numbers make up a "data set." Our two measurements are pressure versus temperature, so each line in the table represents a set of two numbers that go together. If we wanted to express each pair of numbers as (T,P), we have (100 K, 0.200 atm), (200 K, 0.400 atm), etc. Each of these sets of numbers is represented by a particular point in our two-dimensional graph. What we do is draw a dot on the grid for each data set.

Consider the first data point, (100 K, 0.200 atm). Locate 100 K on the temperature axis and 0.200 atm on the pressure axis. (The title of our graph is going to be omitted for sake of clarity.)

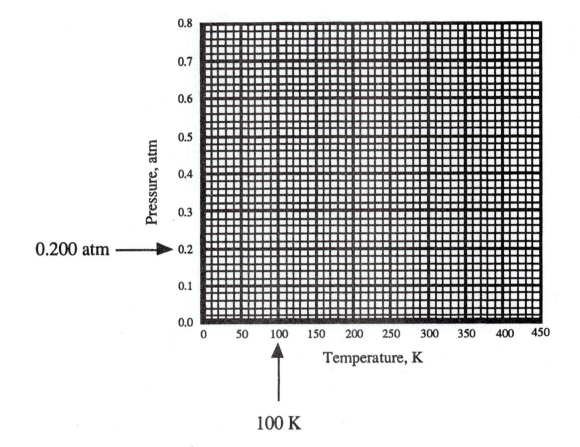

At each value, draw lightly on the graph (or imagine in your mind) a line, starting from each value at the respective axis and into the grid **parallel to the other axis**. The two lines should intersect:

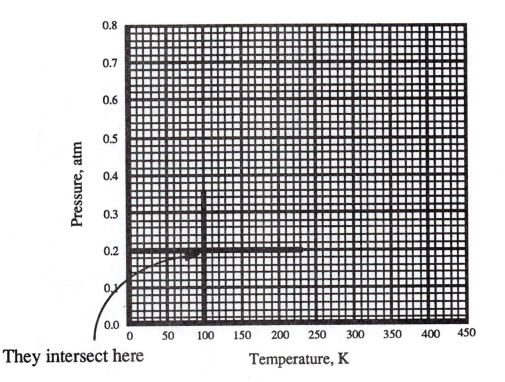

They intersect here

That point of intersection is the point on your graph that represents the data set (100 K, 0.200 atm). Draw a dot at that point:

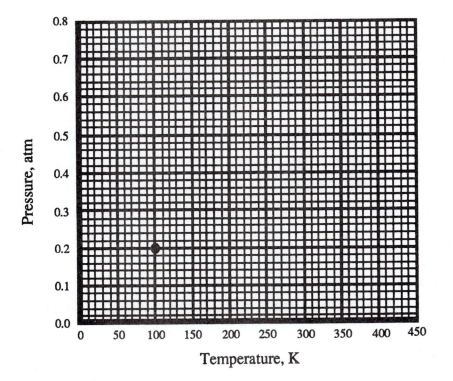

The other five data sets can be plotted in a similar way. When finished, our graph looks as follows:

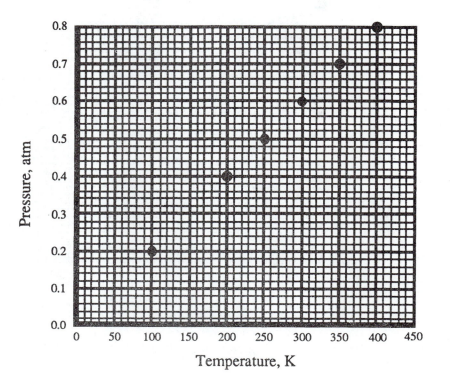

Although the dots used here seem large, they are that size so that they are obvious. You should use as small a dot as you can. Some instructors ask that you circle your tiny dots, so they are more obvious to another person who might be reading the graph.

Drawing the Trend

Remember it was mentioned earlier that almost all graphs illustrate relationships that are <u>smooth</u>. That means that we can establish the overall trend of our measurements with only a limited number of data points, and we can do so with a high degree of confidence. For example, the points that were plotted in the above graph seem to be in a straight line. Even though we do not have measurements for every single point along that line (and we never will, since a line is composed of an infinite number of points), we can confidently predict that if we were to ever measure every single point, all of the points plotted on the graph would make a straight line. In graphing, however, we discover the trend by making only a few measurements, indicated by the dots, and then **smoothly** drawing a straight line through the dots to indicate the overall trend:

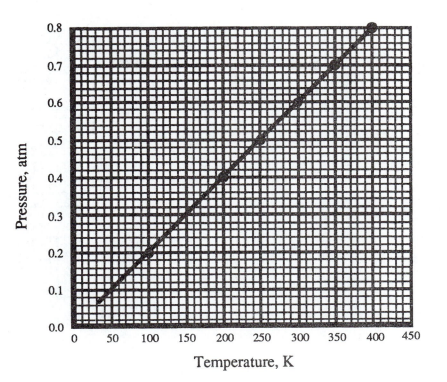

This graph shows that the points are in a straight line. Using this graph, we can understand two things. First, we see that there is a linear relationship between T and P. Second, we can use the graph now to predict the pressure of our gas sample at a temperature **that is not measured experimentally**. For example, what pressure would the gas have at 150 K? The plot shows that "150 K" will intersect with "0.3" atm, and that this is the predicted pressure of the gas.

In reality, the points on the plot won't be this perfect, owing to experimental error. In any real measurement, there will be some uncertainty and some error, and the data set won't be so perfect when plotted. This does not mean that your data are wrong, or that you will not be able to construct a good graph of your data! Remember, you should have an idea of how your data should plot in the beginning, and if you expect a straight line, then the non-ideality of your points may be more a reflection of experimental non-ideality instead of a non-linear (or non-curved) relationship. When a situation like this occurs, how do we draw the graph? We do not simply connect the dots! Instead, we make a "best guess" fit. We draw a line that is a best approximate fit to all of the points, but not a perfect fit to any of the points. For example, if you made temperature and pressure measurements for a real gaseous system and plotted your points, they might have some scatter to them, and a more realistic graph might look as follows:

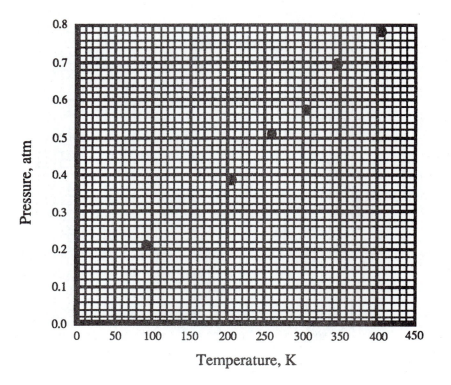

The points are not in a perfect straight line. However, we can "eyeball" a line passing very close to these points, and this line should be a good general representation of the behavior of our system:

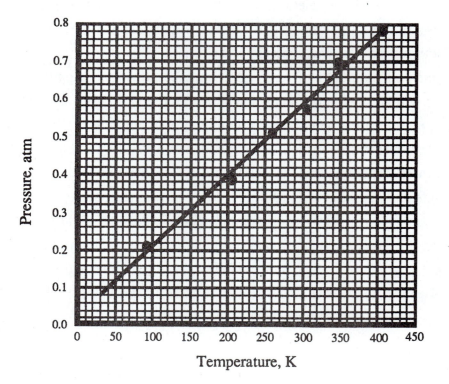

Notice that not all of the points lie directly on the line; some are above and some are below. Usually, about half of the points will be above the line, and about half will be below the line. (This is not always the case, because a "best fit" line also takes into consideration how far the data point is from the line.) This is one example of what would be called a "best fit" line that describes all of these data points.

A true "best fit" line can actually be calculated mathematically, and some calculators can be programmed to determine the mathematically-optimal, true best fit line. However, we won't go into that in this chapter. In most cases in the initial study of chemistry, an "eyeball" best fit line is acceptable. Your instructor should inform you if you are required to use a calculator or computer to calculate a mathematical best fit line for a graph.

Of course, some data will give a curved line, not a straight line. The issues of smooth changes, non-ideal behavior of data, and "best fit" lines also apply. For example, you might have a graph of volume versus pressure that looks like this:

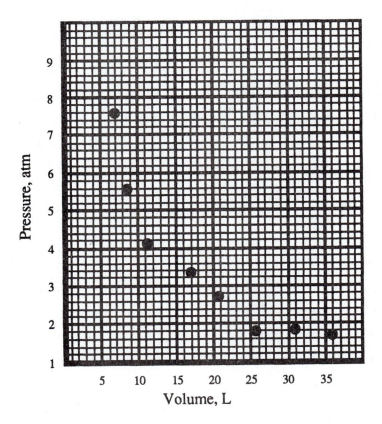

We expect that there is a curved relationship between volume and pressure, and so we can "eyeball" a best fit curved line to represent the smooth change in pressure as the volume changes:

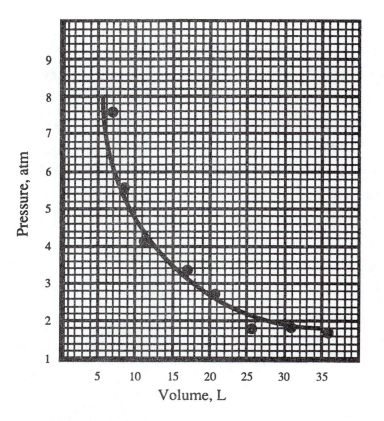

Drawing best fit lines for plotted data in a graph takes practice. In time, you will develop the self-confidence necessary to convince yourself that your ability to guess the best fit line for a set of imperfect experimental data is reasonable. One suggestion is to compare your best fit graphs with the graphs of others using similar data to see how they graph the same information.

Extrapolation

One aspect of graphing involves projecting the line into regions where data do not exist and then predicting values of measurables instead of measuring them directly. This process is called <u>extrapolation</u>. Extrapolation is almost always done with data that are expected to yield a straight line, because it is very easy to extend a straight line as far as you want.

One of the more common uses of extrapolation on graphs in chemistry is to determine the actual value of a y-intercept from data that should make a straight line. For example, in the equation

$$\ln k = \ln A - \frac{E_A}{RT}$$

the "ln A" term is the y-intercept. Because our x axis variable is defined as $\frac{1}{T}$ (see above), we will not get to $\frac{1}{T} = 0$ unless $T = \infty$ (which is physically impossible). Thus, we determine the y-intercept by extrapolation.

The key to extrapolating to a y-intercept is that **the x axis of the graph must go to zero**. Why? Because the true y-intercept of a straight line occurs at x = 0. If x, or whatever expression that is plotted on the x axis, does not have a zero point on the scale, you will not be able to determine the y-intercept by extrapolation.

Consider the following set of data:

k	ln k	T, K	$1/T$, K^{-1}
1.18×10^{-5}	-11.347	293	3.41×10^{-3}
2.35×10^{-5}	-10.659	298	3.36×10^{-3}
4.67×10^{-5}	-9.972	303	3.30×10^{-3}
9.10×10^{-5}	-9.305	308	3.25×10^{-3}
1.81×10^{-4}	-8.617	313	3.19×10^{-3}

If we were going to plot ln k on the y axis versus $\frac{1}{T}$ on the x axis, we would get a graph that looks like this:

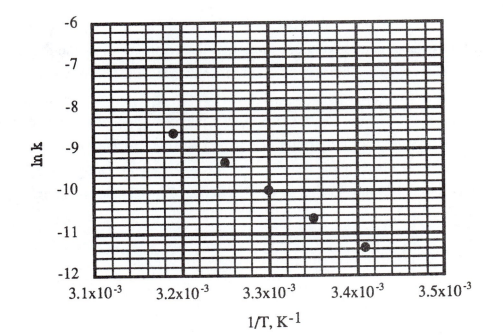

Could we simply draw a best fit line and extrapolate to the edge of the grid to determine the y-intercept? NO! The x axis does not go all the way to zero, which is where the true value of the y-intercept is. If we replot the graph, and make the minimum value of the scale equal to zero, we can extrapolate on that graph to the proper y-intercept. Of course, if we change one scale, we may have to change the other scale so that we can determine an approximate value for the y-intercept. That is the case in this example. But in doing so, we can extrapolate a value for the y-intercept by using these data, and in doing so graphically determine an approximate value for ln A. The graph looks like this:

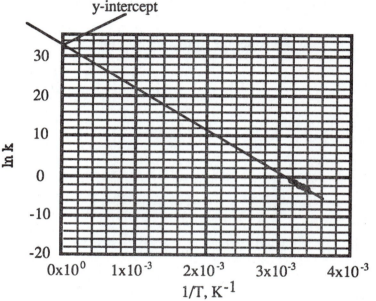

Notice how all of our data are clustered in one section of the graph, and how the plotted points take up only a small portion of the graph. This may be cause for some concern. The relationship between the x-axis variable and the y-axis variable may not be perfectly linear over the entire interval. Large extrapolations like this may be suspect because the exact y-intercept depends not only on the reliability of the original data, but also the appropriateness of our best fit line. If there is too much scatter in the experimental data, or if a poor best fit line is drawn, then extrapolating the straight line for such a distance could induce a major uncertainty in the value of our y-intercept. Consider the two lines below:

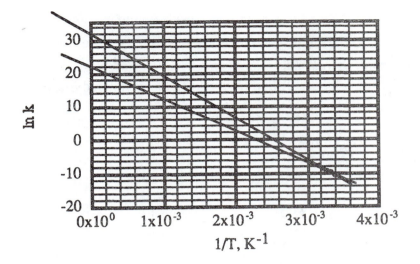

Which line provides the "better" y-intercept? One line gives ln A ≈ 22, while the other gives ln A ≈ 34. Working backwards, these give values of A approximately equal to 3.5 x 10^9 or 5.8 x 10^{14}, respectively. A five order of magnitude difference for a variable is not inconsequential! This difference illustrates the precautions you need to take when performing extrapolations of your data.

Student Exercises

Only a few exercises are given here because there are so many aspects to graphing data and the process is always open to a little personal interpretation. Because of this, no "answers" are given for these exercises. If you wish to check the appropriateness of your graph, ask another student or your instructor to comment on your work for you. Several pages of graph paper are included after the exercises, but you can also purchase graph paper separately and use that.

8.1. Construct a graph of the following data, which relates the density of water in g/cm^3 to the temperature. Draw a best fit curve for the points.

Temperature, °C	Density	Temperature, °C	Density
0	0.9998	50	0.9880
10	0.9997	60	0.9832
20	0.9982	70	0.9778
30	0.9957	80	0.9712
40	0.9922	90	0.9654
		100	0.9584

8.2. Various coefficients are used to represent the non-ideal behavior of real gases. A common one is called the <u>second virial coefficient</u> and has the symbol B. B varies with temperature, and at the temperature where B is zero, the gas acts almost exactly like an ideal gas. Graph B versus temperature for carbon dioxide, CO_2, draw a best fit line to illustrate the trend, and estimate the temperature where CO_2 acts like an ideal gas.

Temperature, K	B
400	-62
500	-30
600	-13
700	-1
800	7
900	12
1000	16
1100	19

8.3.   Hydrocarbons are simple molecules composed of carbon and hydrogen. The simplest hydrocarbons are long chains of carbons with hydrogens attached to the long chain. There is an interesting trend in the melting points of these hydrocarbons: the greater the number of carbon atoms in the chain, the higher the melting point. Use the data in the table below to extrapolate and predict the number of carbon atoms in the molecule that melts around 0°C. Can you look up this information in another reference and check your prediction?

# of C atoms in molecule	Melting point, °C
17	22.0
18	28.2
19	32.1
20	36.8

8.4. A student is making a measurement of the solubility of a salt in a NaCl (sodium chloride) solution. The following data are collected experimentally:

NaCl concentration, M	Solubility of second salt, g/L
0.05	1.7×10^{-6}
0.025	9.5×10^{-7}
0.0125	5.1×10^{-7}
0.00625	2.9×10^{-7}

Graph the solubility versus the NaCl concentration and extrapolate to determine the solubility of the salt in pure water, i.e. where the concentration of NaCl = 0.

Index